NUCLEAR PHYSICS

NUCLEAR PHYSICS

by

W. HEISENBERG

Director of the Max Planck Institute of Physics, Göttingen

PHILOSOPHICAL LIBRARY
NEW YORK

*Published 1953, by the Philosophical Library, Inc.
15 East 40th Street, New York 16, N.Y.*

ALL RIGHTS RESERVED
PRINTED IN GREAT BRITAIN

PREFACE

This book was based originally on a series of lectures, and is intended for readers who, while interested in natural sciences, have no previous training in theoretical physics and yet are familiar to a certain extent with physical ideas. In conformity with the express wish of the *Verband deutscher Elektrotechniker*,* under whose auspices the lectures were given, a short history of atomic physics, as well as a general review of contemporary knowledge of atomic and nuclear structure, are included here as an introduction. Obviously, a thorough understanding of nuclear physics cannot be gained from a short survey of this nature, but it may at least succeed in providing a basis for an understanding of the lectures on nuclear physics which follow. In my treatment of nuclear physics, I have departed somewhat from the method followed by other popular books on the subject, inasmuch as I have attempted to begin my discourse with the theory of the processes and reactions within the atom, and to discuss practical applications in conclusion only. At the same time, it was essential to make the theory intelligible without resort to mathematics, with the aid of illustrative models and by citing as analogies certain more widely known related phenomena. Nuclear physics lends itself to such a treatment more than many other branches of physical science. However, this method obviously has its natural limitations, and for a more profound understanding of the entire complex of relationships, a mathematical presentation of the subject is, of course, essential. For a thorough study of nuclear physics in this sense, there are many excellent books available. In the present volume,

* Association of German Electrical Engineers.

the technical apparatus of nuclear physics is discussed in the seventh chapter only; the eighth, and last, chapter presents a survey of the practical applications achieved up to the present time.

Since the publication, during the war, of the first edition of this book, reports have been published on the great progress in the field of nuclear physics, and especially on those technical developments relating to the atomic nucleus which had till then been restricted to the secret laboratories of the belligerent nations. These new developments are described, in general outline, in the last chapter of this book, where the practical applications of nuclear physics are discussed. Furthermore, those discoveries which were made or published after the war only, are dealt with in the text elsewhere.

The present English edition, appearing some time after the German one, may be of interest in connection with the history and the principles of nuclear physics rather than with respect to its recent development. Since the writing of the book and even since its last revision in 1948 an enormous development of nuclear physics has taken place, in its principles as well as in technical applications. Therefore some of the content of the book may now be commonplace to many readers, some parts are definitely out of date, since new discoveries have changed the picture. In a new edition the shell structure of the nucleus should play a central rôle, since it has simplified our knowledge of the nucleus considerably through the work of Mayer-Göppert, Haxel, Jensen and Suess. In dealing with the nuclear forces one should mention all the new types of mesons that have been found in recent years and their modes of interaction. But it would probably be entirely impossible to give an account of the present state of nuclear physics in a short work. Therefore this book may still serve as an introduction to a field, the knowledge of which would require much more extended studies.

W. HEISENBERG

CONTENTS

1. ATOMIC THEORY, FROM ANTIQUITY TO THE END OF THE NINETEENTH CENTURY *Page*

I   Matter and Atoms in the Philosophy of Antiquity   1
II   Modern Atomic Theory, up to the End of the Nineteenth Century   4

2. MOLECULES AND ATOMS

I   Molecular Structure   15
II   Rutherford's Atom Model   23
III   The Periodic System of Elements   38

3. RADIOACTIVITY AND THE BUILDING BLOCKS OF THE NUCLEUS

I   Radioactivity   42
II   Artificial Nuclear Transmutations   50
III   The Building Blocks of Atomic Nuclei   54

4. THE NORMAL STATES OF ATOMIC NUCLEI

I   The Binding Energy of the Nuclei   66
II   Nuclear Structure   77
III   The Three Types of Nuclear Energy   80

5. THE NUCLEAR FORCES

I   General Properties of the Nuclear Field   90
II   The Nuclear Forces as Exchange Forces   96
III   The Saturation of Nuclear Forces   103
IV   The Stability of Nuclei   104

6. THE NUCLEAR REACTIONS

I   Alpha Radiation   113
II   The Beta Emitters   122
III   Other Types of Spontaneous Nuclear Transmutation   126
IV   Artificially Induced Nuclear Transmutations   127

7. THE TOOLS OF NUCLEAR PHYSICS

I   The Methods of Detection and Observation   141
II   The Procedures for Producing Nuclear Transmutation   149

8. THE PRACTICAL APPLICATIONS OF NUCLEAR PHYSICS

		Page
I	The Exploitation of Atomic Energy for Useful Purposes	159
II	Uranium Fission and Chain Reaction	165
III	The Uranium Reactor	167
IV	Ennoblement of Matter by Nuclear Reactions	172
V	Artificial Radioactive Substances as Tracers	175
VI	Artificial Radioactive Substances in Chemistry	177
VII	Artificial Radioactive Substances in Biology and Biochemistry	182
VIII	Artificial Radioactive Substances in Medicine	184
IX	The Use of Stable Isotopes	186

Appendix: Research in Germany on the Technical Application of Atomic Energy ... 189

TABLES ... 204

INDEX ... 221

ACKNOWLEDGEMENT

Thanks are due to the Editors of *Nature* for permission to reprint the article on page 189.

1. ATOMIC THEORY, FROM ANTIQUITY TO THE END OF THE NINETEENTH CENTURY

I. MATTER AND ATOMS IN THE PHILOSOPHY OF ANTIQUITY

Nuclear physics is one of the most recently developed branches of physics. The term *nucleus* was first introduced by Rutherford about forty years ago, and the more detailed knowledge of the nuclei of atoms is only about fifteen years old. But the concept of the atomic structure of matter—the view that there exist certain smallest, ultimate, indivisible units, which are the basic building blocks of all matter—dates back to the philosophy of Antiquity, and was suggested by Greek philosophers as a daring hypothesis 2,500 years ago. Anybody who desires to understand something of modern atomic theory, will do well to study the history of the concept of the atom in order to become acquainted with the origins of those ideas which now have come to full fruition in modern physics. For this reason, the following lectures, the object of which is a description of the physics of the atomic nucleus, are prefaced by a short survey of the history of atomic theory.

The idea of the smallest, indivisible ultimate building blocks of all matter first came up in connection with the elaboration of the concepts of Matter, Being and Becoming, which characterized the first epoch of Greek philosophy. At the very dawn of ancient philosophy we find a remarkable statement by Thales, who lived in Miletus in the sixth century B.C.: He said that water was the source of all things. As Friedrich Nietzsche expounded, this sentence expresses three of the most essential and fundamental ideas of philosophy. Firstly, the question as to the source of all things; secondly, the demand that this question be answered in conformity with reason, without resort to myths or mysticism—in those times, no idea was regarded as more evident than that the source of all things must be sought in something material, such as water, and not in life—thirdly, the postulate that ultimately, it must be found possible to reduce everything to one principle. Thales' statement was the first expression of the idea of a fundamental substance, from which the whole universe had arisen, although in that age the word *substance* was certainly not interpreted in the purely material sense which we ascribe to it to-day.

In the philosophy of Anaximander, a pupil of Thales, who also lived and taught in Miletus, the idea of a fundamental polarity—the antithesis of Being and Becoming—was substituted for the concept of a single fundamental substance. Anaximander argued that if only one fundamental substance were to exist, this infinite, homogeneous substance would completely fill the universe, and therefore, the great many varieties of phenomena would remain unexplained, and for this reason, Change and Becoming must have arisen from that indeterminate prime basis of all things. Anaximander seems to have regarded the process of Becoming as some sort of degeneration or debasement of this undifferentiated Being—as an escape, as it were, ultimately expiated by a return into that which is without shape or character.

In the philosophy of Heraclitus, the concept of Becoming occupies the foremost place. He regarded that which moves— fire—as the basic element. In the teachings of Parmenides, a fundamental polarity—that of Being and Not-Being—is the central concept. Parmenides, too, regarded the wide variety of phenomena as resulting from the combined action and reaction of two opposed principles.

Anaxagoras, who followed Thales by about a century (he probably lived about 500 B.C.), was responsible for a definite transition to a more materialistic view of the world of phenomena. He assumed that there existed an infinite number of basic substances, the mutual interactions of which produced the variety of world processes. In his view these basic substances possessed the character of purely material elements in

ATOMIC THEORY

a much greater degree; he conceived of them as being eternal and indestructible in themselves, and he considered that the change and sequence of phenomena were produced solely by their sharing in the movement which threw them together at random.

Empedocles, about ten years later, saw the existence of four 'elements'—earth, air, fire and water—as the 'prime root' of all things. He regarded the primordial state of all things as consisting in an undifferentiated, homogeneous mixture of the elements, bound by Love in a state of eternal bliss, whereas Hate tended to separate these elements and to shape out of them the variegated drama of Life.

This pronounced tendency to materialism reached its highest development with the philosophers Leucippus, a contemporary of Empedocles, and Democritus who was Leucippus' pupil. The antithesis of Being and Not-Being became crystallized in the doctrines of Leucippus as the antithesis of 'Full' and 'Empty'. The concept represented by 'Full' was regarded as manifesting itself in the ultimate, indivisible particles, the *atoms*, between which there was nothing but emptiness. The atom was pure Being, eternal and indestructible, but inasmuch as there existed an infinite number of atoms, pure Being could, within certain limits, be repeated an indefinite number of times. Thus, for the first time in history, there was voiced the idea of the existence of smallest ultimate, indivisible particles—the atoms—as the fundamental building blocks of all matter. In this manner, the concept of matter became analysed, in fact, into two sub-concepts: atoms and the void in which the atoms move. Formerly, space had seemed to be filled by matter; it was, as it were, stretched or expanded by material substance, and absolutely empty space had been inconceivable. But now, empty space was allotted a very important function: it became the vehicle for geometry and kinematics, by making possible the various arrangements and movements of atoms.

Although the atom was regarded as having a special position in space, also a shape, and as executing certain movements, it was not allotted any attribute other than these purely geometrical properties. The atom had neither colour nor smell nor taste, and the properties perceptible by human senses, together with their changes and mutations, were supposed to be produced by the movement and displacement of atoms in space. Just as both tragedy and comedy could be written with the same latters of the alphabet, the vast variety of events in the universe were regarded as the products of the selfsame atoms, of their different positions and different motions. Democritus said: 'A thing merely *appears* to have colour; it merely *appears* to be sweet or bitter. Only atoms and empty space have a real existence.'

The basic ideas of atomic theory were taken over and modified, in part, by the later Greek philosophers. Plato, in his dialogue *Timaeus*, co-ordinates these ideas with Pythagoras' theory of the harmony of numbers, and identifies the atoms of the elements—earth, air, fire and water—with the symmetrical bodies, cube, octahedron, tetrahedron and ikosahedron. The Epicureans, too, adopted the essential concepts of the atomic theory, and appended to it an idea which was to play an important part in natural science at a later date: the idea of natural necessity. According to this theory, the atoms are not thrown together arbitrarily, at random, like dice, nor set in motion by forces such as Love or Hate, but their paths are determined by natural laws, or by the working of blind necessity.

After this point, there was no further development in atomic theory, either in the philosophy or in the science of Antiquity.

II. MODERN ATOMIC THEORY, UP TO THE END OF THE NINETEENTH CENTURY

All the progress which we have mentioned, occurred in the course of a few centuries. Two thousand years elapsed then before they were recalled, and before another thinker took up these ideas and transformed them into something fruitful. During the latter part of Antiquity, and during the Middle Ages in particular, the philosophy of Aristotle was accepted as an incontestable foundation, and for the Christian outlook reality had changed to such an extent that the attention of mankind was not attracted by material Nature for a long time.

The first philosopher to revive these neglected trends of thought was the Frenchman Gassendi. A theologian and philosopher, he was born in Provence in 1592 and died in Paris in 1655. He was a contemporary of both Galileo and Kepler, and as such he witnessed the first achievements of a revived natural science. It was about this time, after a barren interval of nearly 2,000 years, that the soil once again became fertile for the progress of scientific knowledge.

The first representatives of this new natural science, including Gassendi, revolted against the authority of Aristotle and turned to other philosophers of the classical era. Thus, Gassendi embraced the teachings of Democritus, which he at once invested with a completely materialistic form. He, too, held that the world was built of ultimate, indivisible units, or atoms, so small as to be invisible. And like Democritus, he regarded the wide variety of phenomena as the product of the variety in the arrangement and movements of atoms. The idea had already suggested itself that physical phenomena could be made intelligible in a much more concrete—one might even say, banal—way with the aid of the atomic theory. Thus, a mixture of water and wine might be compared to a mixture of two different types of sand which has been stirred so thoroughly that the two kinds of grains are completely intermingled, and distributed statistically, by pure chance. The atoms of water and wine would correspond to the grains of sand in their random and indissoluble union. Furthermore, the idea suggested itself that the states of aggregation of matter could likewise be explained by the atomic theory, even though not in the clear and intelligible manner to which we are accustomed in modern times. To-day we know that in 'solid' water—ice—the atoms are packed tightly in ranks and files, as it were. In 'liquid' water, they are also tightly packed, but are in a state of disorder, and move about in this disorderly state. Finally, in water vapour, or steam, the atoms (or more correctly, certain groups of atoms, which we call *molecules*) move in a way which may be likened to a swarm of fruit-flies, at considerable distances from each other.

This idea was taken up by other investigators, and its application to the material world progressed by leaps and bounds. For the Greeks, the conception of atoms was still the means which enabled the world, as a whole, to be understood and which accounted for observable reality. Now it became the means for the understanding of the behaviour of crude, inanimate matter.

The next scientific investigator whom we must mention was an Englishman, Robert Boyle (1627–1691). He was a chemist and physicist rather than a philosopher. His most important work concerned the theory of gases, and he discovered the law that the product of the pressure and volume of a gas at a given temperature is always constant. Chemistry is indebted to Boyle for other important discoveries, too, more especially for the introduction of the concept of the chemical elements in the modern sense. The Greeks had already associated the notion of elements with fundamental natural phenomena—rest and motion, earth and fire—but Boyle associated this notion with chemical processes in a thoroughly materialistic way. Chemistry was able to convert different substances into each other. Boyle's query was: From what substances can the infinite variety of homogeneous substances existing in nature be built up? And furthermore: What are the elements that cannot be resolved any further, and of which all substances are composed, in one way or another? This problem arose out of the originally quite different question raised by the alchemists in the centuries before Boyle. Alchemy had developed out of the fundamental idea that every substance could ultimately be reduced to one basic substance, and that it must be possible, in principle at least, to convert any substance, any type of matter, into any other—mercury into gold, for instance. But all efforts in this direction had always remained futile; such transmutation could never be effected by chemical means. It appeared obvious that matter was not homogeneous in this sense—when treated by chemical means—but there had to exist basic substances which no chemical process could change into another. Since Boyle's time it has become a matter of common knowledge that there exists a whole series of these basic substances, or *chemical elements*, as against the approximately half a million uniform *chemical compounds* known to-day. The number of chemical compounds exceeds by far that of the basic elements. Nevertheless, the number of the elements is still large enough to make it difficult for us to regard them as the ultimate, indivisible building blocks of matter. Of course, Boyle knew relatively only few of the ninety-six elements known to us to-day, but nevertheless he succeeded in formulating quite clearly the aims and tasks of chemistry. He said: 'What we have to do is to determine into what basic substances matter can be analysed by chemical means, and what these basic substances are.' Thus we see that his chemical elements had nothing more in common with earth, air, fire and water, the elements of Democritus.

A century later came Lavoisier, the real father of modern chemistry. He was born in 1743 and died, a victim of the French Revolution, in 1794. His permanent contribution to science was the founding of quantitative chemistry. He was first to interpret rightly the process of combustion. Up to his time, it had been believed that in the combustion of any body a substance called *phlogiston* was released, and therefore, bodies would necessarily become lighter after combustion. Lavoisier adopted the opposite view, that combustion consisted in the combination of an element with oxygen, and as a result, the body must become heavier. His theory was proved correct by experiment. At the same time, he accomplished something of vast importance, in that he stimulated chemists to investigate changes in mass due to chemical changes.

Now we come to a law which was formulated by Lavoisier in 1774, but became the common property of chemists only several years later: the law of the conservation of mass. Lavoisier already claimed that in every chemical change the total mass of the substances involved remains constant—meaning that the total quantity of converted matter weighs exactly as much after the conversion as it did before it. The discovery and formulation of this law marks the actual beginning of modern chemistry, and in a very few years it became the connecting link between the chemistry of Boyle and the atomic theory of Gassendi.

In 1792, the German Richter discovered that chemical elements always combine in chemical compounds in definite quantitative proportions. It is not possible for just any arbitrary quantity of hydrogen to combine with just any arbitrary quantity of oxygen to form water: hydrogen and oxygen must always combine in the proportion of 1:8 to form water. Otherwise, there remains an unconverted residue of hydrogen or oxygen. This *law of constant proportions* was then raised by Dalton to the status of a fundamental law of chemistry, and within a fairly short time it led to the union of chemistry with atomic theory. Dalton stated this law in a more precise form, and gave it a geometrical interpretation.

It is this very geometrical interpretation that is of paramount importance. We shall try to make it more intelligible by an example. When hydrogen combines with oxygen to form water, we must visualize this process as a mutual combination of the smallest particles—the atoms—of both elements in a higher, more complex unit, a water molecule, according to our modern terminology. We are now in a position to visualize a molecule as a geometrical structure composed of individual atoms, and a water molecule as a structure composed of two hydrogen atoms and one oxygen atom. This view makes the law of multiple proportions directly understandable. The compound which we call water is thus characterized by the 1:2 ratio of oxygen and hydrogen atoms.

This theory of Dalton, advanced in 1803, of atoms combining in molecules in a manner capable of geometrical illustration, was developed further and raised, within a few years, to the status of an established scientific postulate. As early as 1811, Avogadro announced a daring hypothesis, by which he laid the corner stone of what we know to-day as the chemical theory of atoms. He maintained that at the same temperature and pressure, equal volumes of all gases contained the same number of molecules. Although this hypothesis was still in need of experimental proof, it soon turned out to be the clue to the determination of atomic weights, and it also provided a solid, permanent foundation for Dalton's atomic theory. If the number of atoms or molecules contained in a specific quantity of gas is known, the composition of an individual molecule can be stated exactly—for instance, whether a water molecule actually consists of one oxygen atom and two hydrogen atoms.

Thus the way was paved for a quantitative determination of the weight or mass ratios of atoms. Although the absolute number of atoms present at any one time was not known, it was known for a certainty that at the same temperature and pressure, the number of molecules contained in equal volumes of gases was the same. This was sufficient, since it furnished information concerning the mass ratios of atoms and molecules.

Not much later, the Swedish scientist Berzelius determined the atomic weights of a great many molecules, and succeeded in developing very definite theories about the way in which molecules are built from individual atoms. Berzelius also studied the nature of the forces binding atoms together in molecules. It was he who introduced the notion of *valency* in connection with the force binding an atom of one element to an atom of another. In studying this force, he came to the conclusion that it must be of electrical nature.

The status of atomic theory about 120 years ago can be summed up as follows: It was known that the prodigious number of chemical compounds could be reduced to a relatively small number of chemical elements, a great many—although by far not all—of which were known. The mass ratios of the atoms of these elements were also known fairly accurately—as, for instance, that one oxygen atom was roughly 16 times, and a nitrogen atom roughly 14 times heavier than a hydrogen atom. However, there were still considerable gaps to be filled. Nothing at all was known about the absolute size of atoms, or the order of magnitude of their number within a given volume of space. All that was known was that in gases at the same temperature and pressure, the number of molecules was the same. So far as accurate knowledge was concerned, an atom might still, as Democritus had believed, be approximately the same size as one of the motes dancing in a sunbeam, or infinitely smaller. Just as little was known about the shape of atoms, or about the forces operating between them. As to the latter question, nothing but extremely hypothetical conjectures was possible. Furthermore, although it was known that, chemically, the atoms must be the ultimate building blocks of matter—in other words, the smallest units demonstrable by chemical means and methods—no one knew whether or not these chemical atoms might be capable of being further subdivided and transmuted into each other by the application of other methods.

A discovery, from which Prout was the first to draw conclusions in 1815, actually spoke against the theory of the absolute indivisibility of atoms. Prout (1785–1850) based his deductions on the fact that the atomic weights known in those days—these were mainly only those of the lighter elements—were nearly integral multiples of the atomic weight of hydrogen. This fact is the basis of his view, that all atoms were built up of hydrogen atoms. Since one carbon atom was roughly twelve times, and one oxygen atom roughly sixteen times as heavy as one hydrogen atom, the carbon atom had to be composed of twelve hydrogen atoms, and the oxygen atoms of sixteen hydrogen atoms. The hydrogen atom would thus be the only, ultimate building block of all matter. The hypothesis which postulated the existence of nearly a hundred different elements, had always been regarded as rather difficult to accept. For if we really believe in homogeneity in nature, we must obviously prefer the number of basic elements to be considerably smaller.

Despite the captivating features of Prout's hypothesis, it remained almost completely neglected for more than a hundred years. The main reason for its being discarded was that the atomic weights of the heavier atoms had not been proved to be approximately whole numbers. Nevertheless, this hypothesis did actually contain a grain of truth of considerable importance. We shall see later on that, in a modified form, it plays a fundamental part in modern nuclear physics.

A new era in atomic theory was introduced by Faraday (1791–1867), who combined atomic theory with the theory of electricity. Atomic theory is indebted to him for the formulation of a law of prime importance: In chemical changes induced electrically—by electrolysis—there is always a relationship between the matter transformed and a definite quantity of electricity. Moreover, Faraday discovered that the masses of the substances transformed by a certain definite quantity of electricity, are related to the so-called 'equivalent weights,' and thus in the simplest cases—those of univalent substances— to the atomic weights of the substances concerned. This discovery indicated that electricity, too, possessed an atomic structure, of such nature that each atom, or each molecule, of a chemical compound was always associated with one or more atoms of electricity, even though in some manner then still unknown. This was pointed out by Weber as early as 1848.

It explains logically why the same quantity of electricity was always associated with the same quantity—in other words, the same number of atoms—of a substance. To-day, the *mol* (also written *mole*) and the *gramme-atom* are customarily employed as basic mass units. One mol of any substance is that quantity which weighs as many grammes as the numerical value of its molecular weight; one gramme-atom of an element is a quantity equal in grammes to the numerical value of its atomic weight. Thus one mol of oxygen gas, O_2, (molecular weight, 32) is equal to 32 grammes of oxygen gas, and one gramme-atom of oxygen, O, (atomic weight, 16) is equal to 16 grammes of oxygen.* Every gramme-atom of a univalent element is associated with a certain quantity of electricity, 96,520 coulombs, and one gramme-atom of a multivalent element is associated with the appropriate multiple of this quantity.

The next advances were made in the domain of the theory of gases, raised to the status of an exact science as a result of the efforts of Maxwell, Boltzmann and, above all, Clausius. It was through the work of these investigators that the concept of a gas as something consisting of molecules in rapid motion—comparable in a way to a swarm of midges—gained a solid foundation in conformity with strict mathematical principles.

The year 1865 brought a new achievement of considerable importance: The first, and as yet only approximate, determination by Loschmidt of the size of atoms, and hence of the number of molecules contained in a given volume. Like Robert Mayer before him, Loschmidt investigated the internal friction of gases, and as a result of the preliminary researches of his precursor, he obtained a first clue to the size of an atom. His conclusions were still rather inaccurate, but he arrived at the right order of magnitude. The size of the atom has been known accurately for about forty years only. To give some idea of it, let it be mentioned that approximately

---

* According to the physical scale of atomic weights, where the atomic weight of the oxygen isotope $_8O^{16}$ is 16·0000.

ten million atoms placed next to each other in a straight line would form a line of one millimetre in length. Therefore, individual atoms are completely invisible, and it is impossible to observe them directly. They are infinitely smaller than those motes in a sunbeam which Democritus regarded as comparable with atoms with respect to order of magnitude.

The following years produced another step forward in the field of our knowledge of electricity. Through Faraday's discoveries, the existence of atoms of electricity had become a probability, but they were still known solely in association with atoms of the chemical elements, and not in a free state. Free atoms of electricity, atoms not bound to atoms of ordinary matter, were discovered by Hittorf in cathode rays, which are a consequence of electrical discharges in highly attenuated gases. Hittorf (1824–1914) studied the deflection of cathode rays in magnetic fields, and found that on the ground of the magnitude of this deflection it was possible to calculate the ratio of the charge to the mass of those particles moving in the cathode rays. As the mass of an individual atom had been known since Loschmidt's time, and as on the ground of Faraday's discoveries the size of the atoms of electricity could be computed to a fair approximation, the magnitude of the mass with which the atom of electricity was associated in its free state— in the cathode rays—was now determined in connection with the ratio just mentioned. As a result of more recent measurements, it is now known that this mass is about one 1,840th part of the mass of the lightest of all atoms, the hydrogen atom.

These atoms of free electricity are called *electrons*, a name first suggested by Stoney.

Of major importance is the fact that the wide range of masses like those of the atoms of the chemical elements is absent in the case of the atoms of electricity. Electrons are always associated with the same mass, which circumstance is in excellent conformity with the demand for homogeneity in nature.

The view that electrons might be, in some way or another, constituent parts of atoms, gradually developed during the years that followed. An amazing fact was that only negative electricity could be observed in a free state, as electrons, while positive electricity would always appear in association with atoms of matter. This fact of experience indicated that atoms contained negative electrons as component parts, and therefore, free negative electricity could manifest itself only when an electron was torn off an atom, with the result that an equivalent quantity of positive electricity remained bound to the remainder of the atom. But fifty years ago it was totally impossible to arrive at a clear notion of these phenomena. The weights of atoms were approximately known, and so was the volume occupied by them. It was known also that atoms possessed electrical properties, and that they contained one or more electrons. But little or nothing at all was known about their structure, and the question as to their shape could not even be asked.

The solution of this problem was reserved for the twentieth century, the threshold of which we are now approaching in our survey of the history of atomic theory. The further course of this history is so closely linked with the topic proper of this book that in the following chapters it may be presented in connection with the latter only.

2. MOLECULES AND ATOMS

I. MOLECULAR STRUCTURE

In order to prepare the way for the actual subject of this book, the second lecture will deal first with the structure of molecules, and then with the structure of atoms.

Visualize a piece of silver. You can cut it up, first, with coarse mechanical tools, and then with a file you can reduce these fragments to almost invisibly minute specks of dust. But neither of these operations has enabled you to reach the smallest component particles. If you rub the silver with your hand, a minute quantity of the metal will stick to your hand. But even this hardly at all visible quantity contains an enormous number of silver atoms. Finally, you can heat the silver until it melts and finally evaporates, in other words, until it becomes a gas. This process breaks it up into its ultimate, smallest particles, the atoms. At any rate, it cannot be broken up further either mechanically or by chemical means. Silver is a pure element.

But if a drop of water is caused to vaporize, it is not broken down into water atoms. The smallest particles which can be obtained in this manner, the water molecules, can still be broken down further by chemical means. A water molecule consists of two hydrogen atoms and one oxygen atom. Water is not an element.

Modern physics has established a definite geometrical picture of the structure of such a molecule. The hydrogen atom is denoted by the symbol H, the oxygen atom by the symbol O. With these symbols, the water molecule can be represented by the following schematic formula:

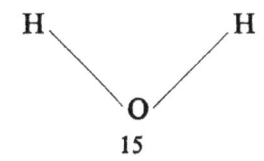

$$H \diagdown_{O} \diagup H$$

For certain reasons, which cannot be discussed further here, the water molecule is imagined as the triangular structure shown diagrammatically in Figure 1. In this sketch, shading indicates the average distribution of the electric charge of the atoms.

Now, proceeding from the individual molecule to matter in its various states of aggregation—for instance, from the water molecule to water vapour, water and ice—we can extend this visual presentation as follows: In water vapour, the molecules

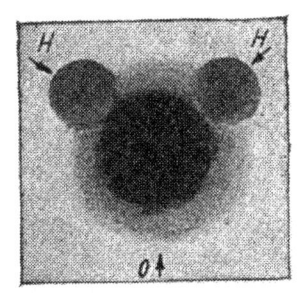

Figure 1.—Model of the molecule of water.

dart about erratically, in all directions, like a swarm of midges, at considerable distances from each other, and in a state of complete disorder. Their movement is closely dependent upon the temperature of the vapour; heat is always associated with a disorderly motion of the molecules. In the case of the larger particles, a powerful microscope will enable us to discern the so-called *Brownian movement*, which becomes always more and more marked as the temperature of the substance increases. In liquids, the molecules are likewise in a state of disorder, but they are closely packed and move in between each other, in a manner comparable to the movement of ants in an ant-hill. In crystalline, solid matter, the atoms, or molecules, are likewise tightly packed, but here they constitute a completely orderly pattern. Figure 2 shows the model of a rock salt crystal. Rock salt is a chemical compound of the elements chlorine and sodium. In this model, the black dots stand for chlorine atoms, the white dots for sodium atoms. They alternate in the crystal in a perfectly regular pattern. Actually, these atoms are in a state of more or less violent movement, according to the temperature of the crystal; they vibrate about their positions of equilibrium. Since the atoms are actually packed tightly, without any empty space between them, the above model does not exactly correspond to reality.

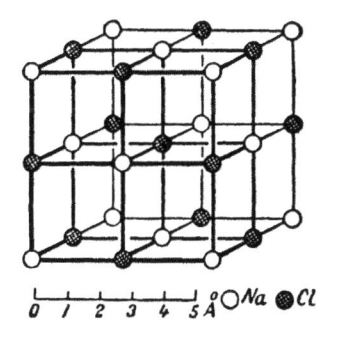

Figure 2.—Model of a common salt crystal.

The question now is: What do such schematic models actually mean, and have we no reason for viewing them with a certain mistrust? For if atoms are supposed to be the smallest units of matter, they cannot be expected to behave in every respect like the visually perceptible objects of our daily experience—like the actual black and white dots in the spatial model of a crystal, for instance. One is indeed quite justified in suspecting that here, in a region where we are approaching the ultimate, fundamental component parts of matter, there is a limit to our power of perception, too. Hence, we must ask, first of all, what the actual size of the atoms is, and what magnification would be required to make a molecule appear to us the size, say, of a billiard ball. Secondly, we must ask, to what extent such a visual model can be justified at all—in other words, what its intended purpose is. Does it possess such a directly perceptible import that we may expect an ideally perfect microscope of the future actually to show us such an image of a real molecule?

Let us take first the question of the size of atoms. Obviously, not all atoms are the same size, but all atoms are more or less of the same order of magnitude. The magnification necessary to permit us to see an atom as a structure with a diameter of, roughly, 10 centimetres, is approximately that which we would have to use to magnify a ball of 1 centimetre in diameter to the size of the earth. This example will give you an approximate idea of the infinitesimal smallness of molecules.

Now, the second question, that of the significance of a molecular model: In recent years, a quite novel microscope has been developed, the *electron microscope*, which—unlike the ordinary optical microscope—does not utilize light rays, but electron rays. With this electron microscope, a considerably greater resolving power, and a far greater magnification, can be attained than with the optical microscope, so that it enables us, even at the present stage of its development, to see particularly large molecules as discrete particles. If it should ever become possible to increase this magnifying power to twenty or thirty times its present degree—to be sure, a difficult problem—an individual water molecule may conceivably be made visible through such a microscope.

But the important question is whether we would then see anything in any way resembling the model shown in Figure 1. It is true, of course, that no molecule is ever in a state of rest. It moves, as a whole, under the influence of the temperature, and its component parts vibrate, in reciprocal movement. Thus it would be necessary to take cinematographic records of molecules, in which case one would actually obtain a snapshot of the kind shown in Figure 1. This cannot be doubted in the light of all our present knowledge of atomic physics—and this realization is at the same time a recognition of the visual-perceptual significance of models of the kind shown in Figure 1. However, as a result of thermal movement, minor changes would be constantly taking place in this picture.

How does the physicist know that a water molecule consists just exactly of two hydrogen atoms and one oxygen atom, instead of—let us say—four hydrogen and two oxygen atoms, which would still represent the same mass ratio? In order to answer this question, we must make reference again to the theory of gases, and to Avogadro's hypothesis in particular, which states—as mentioned above—that at the same temperature and pressure, equal volumes of all gases contain the same number of molecules. There exists an exact proof of this hypothesis, but here we shall confine ourselves merely to elucidating it. The pressure on the sides of a vessel filled with a gas, is due to the impacts of the gas molecules, which hammer against those sides like drops of a steady rain, only to be bounced back from them again. The sum total of the forces of these impacts exerts a pressure against the sides. The magnitude of this pressure, obviously, depends on the kinetic energy of the molecules, which, in turn, depends on the temperature of the gas. Since the days of Maxwell, we have known that at the same temperature the molecules of all gases always have the same average kinetic energy. Thus, if equal volumes of gases contain the same number of molecules, it follows that at the same temperature their pressure, too, must be the same. This is the real meaning of Avogadro's hypothesis.

Moreover, at this point, we must refer to certain chemical facts. The fact, already mentioned, that 2 grammes of gaseous hydrogen and 16 grammes of gaseous oxygen combine to form 18 grammes of water vapour, can be expressed by the following equation:

2 gm. of hydrogen + 16 gm. of oxygen = 18 gm. of water vapour.

Instead of the ratio by mass, we may investigate also the ratio by volume present when hydrogen and oxygen combine at the same temperature. Experiments show, for instance, that 1 litre of hydrogen and ½ litre of oxygen combine to form 1 litre of water vapour. Thus, we get the following equation:

1 litre of hydrogen + ½ litre of oxygen = 1 litre of water vapour.

From these two equations, it is easy to deduce the mass ratios of these three kinds of molecules. Since 1 litre always contains the same number of gas molecules, all we have to do is determine the ratio of the above-mentioned masses and volumes of the three gases, to obtain numbers proportionate to the masses of their molecules. Thus we get:

Hydrogen gas, 2 gm. litre;
Oxygen gas, 32 gm. litre;
Water vapour, 18 gm. litre.

Accordingly, the relative masses of the various molecules in question are:

Hydrogen gas: oxygen gas: water vapour = 2: 32: 18

The hypothesis which expresses these facts most simply, and which has been confirmed by other experiments, is that a hydrogen molecule consists of two hydrogen atoms of atomic weight 1, and consequently, has the molecular weight 2. Similarly, an oxygen molecule consists of two oxygen atoms, of atomic weight 16, and consequently, has the molecular weight 32. And, finally, a water molecule consists of two hydrogen atoms and one oxygen atom, and consequently, has the molecular weight $2 \times 1 + 16 = 18$. This compound can be described by the following formula:

$$1 \ H_2 + \tfrac{1}{2} O_2 = 1 \ H_2O$$

The atomic weights of all elements, as well as the molecular weights of all chemical compounds, can be determined in analogous manner. The unit of atomic weight is chosen so as to be equal exactly to $\frac{1}{16}$ of the chemical atomic weight of oxygen, so that the atomic weight of oxygen is exactly 16·0000. Similarly, the molecular weight is the standard measure of the mass of a molecule measured in terms of the same unit.

We have already mentioned the unit called *mol* (or *mole*). One mol represents that quantity of a substance which weighs as many grammes as equal numerically the molecular weight of that substance. Thus, one mol of hydrogen gas (H_2) is 2 grammes, one mol of oxygen gas (O_2) is 32 grammes, and 1 mol of water (H_2O) is 18 grammes. Thus the masses of 1 mol of different substances are to each other exactly as their respective molecular weights, and consequently, as the masses of their individual molecules. It follows, therefore, that 1 mol of any substance always contains exactly the same number of molecules. Quite analogously, a gramme-atom is defined as that quantity of an element which weighs as many grammes as equal numerically the atomic weight of that element. Thus 1 gramme-atom of hydrogen (H) is 1 gramme, one gramme-atom of oxygen (O) is 16 grammes. It is evident that 1 gramme-atom of any element contains always the same number of atoms—namely, as many atoms as there are molecules in 1 mol. A knowledge of these concepts is very important, for they enable us to count molecules, or atoms, by weighing them, as it were, and for this reason it is absolutely necessary to know the exact number of molecules contained in a mol. As we have stated before, this number was first computed correctly—as regards the order of magnitude at least—by Loschmidt in 1865. But the first really reliable calculation was made in 1900 only, on the basis of Planck's radiation law. To-day, the most reliable numerical value of this important Loschmidt's number (L) is: $6·024 \times 10^{23}$.

This means that 1 mol of a substance—for instance, 32 grammes of oxygen gas—contains almost 1 quadrillion molecules.

Furthermore, Loschmidt's number gives us an exact knowledge of the masses of the individual atoms and molecules in terms of the customary unit of 1 gramme. Since 1 mol of hydrogen gas weighs 2 grammes, we can easily find—dividing by Loschmidt's number—that the mass of a hydrogen molecule (H_2) is $3 \cdot 34 \times 10^{-24}$ gramme, and therefore, the mass of a hydrogen atom (H) is $1 \cdot 67 \times 10^{-24}$ gramme. The masses of all other atoms and of all kinds of molecules (when their atomic structure is known) can be computed.

Now let us turn our attention to the magnitude of the electric charge which is linked, through electrolysis, with an atom or molecule—the atom of electricity, the electron. We have already stated that a certain quantity of electricity, namely:

$$F = 96{,}520 \text{ coulombs}$$

is always linked to 1 gramme-atom of a univalent substance. But it is very difficult to appreciate the magnitude of this charge. It is much greater than any electric charge that can be produced in any single substance in the laboratory. If both the moon and the earth were to carry, each, a charge of this magnitude, they would attract or repel each other—despite the great distance between them—with a force equivalent to several hundreds of kilogrammes. This is the charge carried by 1 gramme-atom of a univalent substance. But since 1 gramme-atom always contains the same number of atoms (Loschmidt's number), the charge carried by an individual univalent atom is determined by dividing the equivalent charge F by Loschmidt's number; this charge is $e = 1 \cdot 6 \times 10^{-19}$ coulombs, or $4 \cdot 8 \times 10^{-10}$ electrostatic units, and is therefore exceedingly small. This charge of an atom of electricity is called the *elementary quantum of electricity*, for any electric charge can be only an integral multiple—either positive or negative—thereof.* This number and several other important

* *Translator's note:* This elementary quantum of electricity (*elektrisches elementar Quantum*) is more frequently called *electronic charge* in English-speaking countries. The former term is, however, used throughout the present book.

constants of nuclear physics, are given in Table 1 at the end of this book.

We have already pointed out that the ratio of the charge to the mass of electrons, both in electric and magnetic fields, was determined on the basis of the deflection of the cathode rays—of electrons, in other words. Once the magnitude of their charge is known, the electronic mass can be computed. It is only about 1/1,840 of the mass of the hydrogen atom, viz.: $9 \cdot 1 \times 10^{-28}$ gramme. In Table 1 it is designated *rest mass*, because the mass of every body increases with its velocity.

Up till not so very long ago, only electrons with a negative electric charge were known. Positively charged electrons were only discovered during the past decade. Under normal conditions, this positive electron is very short lived; as a rule, it vanishes soon after it comes into being. Otherwise a positive charge also occurs invariably in magnitudes equal to one or more elementary quanta of electricity, in association with masses of the atomic order of magnitude. This fact in itself suggests that the mass of the atom is associated with a positive electric charge which is neutralized by negative electrons, and that ions are produced either by a loss or gain of electrons. But it was still a long way from this concept to the creation of the correct atom model.

II. RUTHERFORD'S ATOM MODEL

Shortly before the end of the nineteenth century, the way was opened for new decisive developments in nuclear physics. They were introduced by a discovery not directly related to nuclear physics at all: the discovery of x-rays by Wilhelm Röntgen in 1895. The first effect of this discovery was merely the knowledge of a new type of radiation which, although not perceptible directly by human sense organs, could be measured by instruments in the physical laboratory, and which amazed the entire world by its power to penetrate dense layers of matter.

In the following year, as a result of his attempt to discover other rays of a similar nature, the Frenchman Henri Becquerel proved that certain substances, notably the compounds of uranium, emitted certain rays of a similar penetrating power, and that this emission took place independently of any external agency. This phenomenon became known as *radioactivity*, and the entire modern theory of atoms owes its development to this discovery. A series of other important developments then followed in rapid succession. The two Curies succeeded as early as 1898 in obtaining from uranium a much more strongly radioactive substance, and because of its extremely strong radioactive properties they named it *radium*—that which radiates.

Just about that time, Ernest Rutherford, the real father of modern atomic physics, began to take a hand in its development. He was born in Nelson, New Zealand, in 1871, and died in Cambridge, England, in 1937. Shortly after the announcement of the phenomenon of radioactivity, he found that rays of various types were emitted by radioactive substances, and these rays were distinguished by their different absorbabilities in matter. They are called *alpha, beta* and *gamma rays*. The alpha and beta rays are deflected by a magnetic field, which fact indicates that they carry an electric charge—the alpha ray a positive charge, the beta ray a negative one. Gamma rays cannot be deflected—they carry no electric charge.

A more precise study of the alpha rays led to the conclusion that they consisted of particles in rapid movement, each carrying two elementary quanta of positive electricity, and having a mass equal to that of a helium atom (atomic weight 4). The particles which constitute the beta radiation carry only one elementary quantum of negative electricity each, and the mass of each such particle is equal to that of an electron. In other words, beta rays consist of negative electrons in rapid movement. Finally, the gamma rays are identical with x-rays as regards their general properties.

An excellent method was developed by Wilson to make these rays visually detectable. Air containing saturated aqueous vapour is caused to expand suddenly within a chamber, called a *cloud chamber*; it is cooled in consequence and the water vapour therefore becomes supersaturated and has a tendency to condense; though it may remain supersaturated for some time. If now at this moment an alpha particle passes through the chamber, it will tear electrons off the atoms of air along its entire path, leaving positively charged air molecules in its wake. These positively charged air molecules are called *positive ions*. These ions further the condensation of the supersaturated water vapour; in fact, they form 'condensation nuclei', around which the water vapour condenses to small droplets. Thus, all along the path of the particle, a very fine track of water droplets is formed, similar to the condensation bands left behind by an aeroplane at high altitudes, thus producing a picture of its path. This phenomenon is shown in Figures 3 (*a*) and (*b*).

The radioactive preparation is under the visible section of the cloud chamber. The tracks of the individual alpha particles can easily be seen and constitute groups of approximately the same length. In Figure 3 (*a*), two groups of different 'ranges' are easily discernible.

One of the tracks in Figure 3(*b*) shows a sharp deflection. Obviously, something out of the ordinary has happened here to the alpha particle. It has come into the vicinity of the nucleus of an atom and been deflected by it. But, as a rule, the particles move in a straight line, and their range varies mostly between 2 and 10 centimetres.

It is very remarkable that the alpha particles are capable of covering such great distances in a straight line. For it is easy to calculate that on their way they must encounter an enormous number of atoms. This is an obvious inference from the large number of water droplets formed. It seems therefore as if the atoms were not impermeable to particles of such small size, as if, in fact, the latter were able to penetrate the atoms without hindrance. Evidently, they seldom encounter any serious obstacle on their way; but when they do, their tracks will show such deflections, as the one which we see in Figure 3 (b).

At an even earlier date, Lenard had investigated the passage of rapidly moving electrons through matter, and discovered

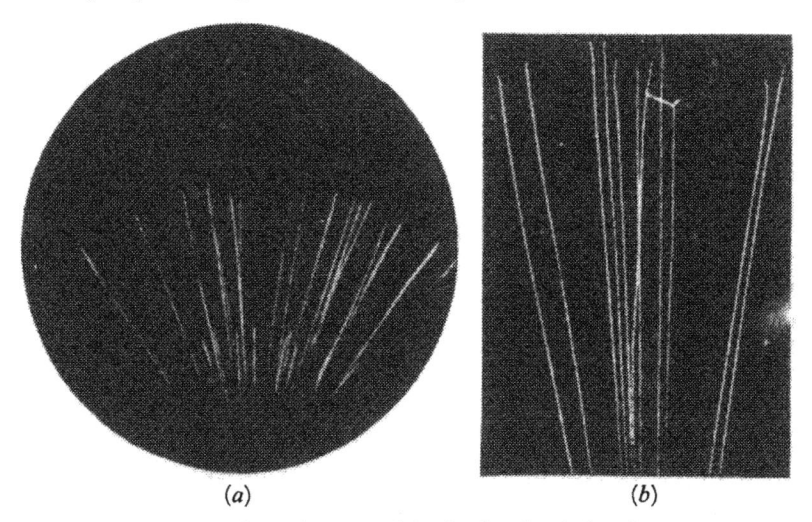

Figure 3.—Alpha particles in the cloud chamber.

that electrons were capable of penetrating extraordinarily thick layers of matter. Thus he reached the conclusion that the space occupied by an atom is, mostly, empty, and the path of an electron is influenced by individual centres of force only, which he called *dynamides*. It was Rutherford who, as the result of similar studies, took the important step which eventually led to the construction of the first atom model. He studied the tracks of alpha particles through thin metal foil, and concluded, on the ground of the extreme rarity of any notable deflection, that only a very small part of the atom offered any resistance at all to the alpha particles, and that this very small part contained practically all of the mass of the atom. If this were not the case, the laws of elastic impacts would make the occurrence of those considerable deflections, which could at times be observed, quite impossible. On the basis of his observations, Rutherford was able to calculate the volume of the space actually occupied by mass, and concluded that all the rest of the space occupied by the atom must—for all practical purposes—be empty. Rutherford's collaborators, Geiger and Marsden, succeeded in establishing, furthermore, that the deflections of the positively charged alpha particles were produced by electric forces, due to a likewise positive charge on the central part of the atom. Evidently, this central part of the atom and the alpha particles mutually repelled each other, in conformity with Coulomb's well-known law.

These findings were the foundation on which Rutherford built the following atom model: An atom consists of a positively charged nucleus, which contains practically all of the mass of the atom, but occupies only a small fraction of the total volume of the atom as a whole. The positive charge of the nucleus is offset by electrons which, held captive by the attraction exerted by the nucleus, revolve around the nucleus at relatively great distances. These electrons constitute the extranuclear structure of the atom. The number of revolving *planetary* electrons is determined by the nuclear charge. Since each electron carries one elementary charge of negative electricity, the number of electrons must equal the number of elementary charges of positive electricity carried by the nucleus, for the atom as a whole to be electrically neutral. Evidently, the number of electrons determines all the external properties of the atom, and thus, in particular, the forces which it is capable of exerting on other atoms—in other words, its chemical properties, which, too, are ultimately determined by the nuclear charge.

In order to obtain some idea of the orders of magnitude within an atom, imagine an atom, including its extranuclear electron structure, as a sphere with a diameter of approximately 10 centimetres. In such a model, it would hardly be possible to represent the electrons and the nucleus in their proper porportions; they are too small to allow us to do so. In such a model, the nucleus would appear as a minute particle of dust about 1/100 millimetre in diameter, and the electrons would be more or less the same size.

The lightest of all atoms, the hydrogen atom, consists of a nucleus carrying one elementary quantum of electricity, and is therefore circled by one single electron. In this case, the nucleus is said to carry the unit charge or to have the charge number 1.

If we want to obtain a visual model of this atom, we must remember that the electron must move in a path around the nucleus, like a planet around the sun, for otherwise it would fall into the nucleus. We must therefore draw the path of the electron, and we shall assume for the moment, as a working hypothesis, that it is a circular orbit.

Figure 4.—Model of the hydrogen atom.

The hydrogen atom is the simplest of all atoms. The next one is the helium atom with its charge number 2—two elementary quanta of electricity on its nucleus and two planetary electrons in its extranuclear structure. And so forth, up to the heaviest atom known at the present time, curium, which has the nuclear charge number 96. All the necessary information on this subject is given in Table II, at the end of this book.

Once again, there arises the question, to what extent the visual representation supplied by Rutherford's atom model may be accepted literally, at its face value. Can we expect that some day, with the aid of a supermicroscope, we shall actually be able to see electrons revolving in their orbits around a nucleus? In view of the movements of the electrons, we would have to take snapshots. According to our present knowledge, we cannot very well doubt that a snapshot of a hydrogen atom would show actually such a picture as we have described: two point charges at a distance of a few ten-millionths of a millimetre. This is the practical significance of Rutherford's atom model. Obviously, this would not be a picture in definite colours, as our photographs would not be taken with rays of visible light, but with electron rays. But in any case such a snapshot of a hydrogen atom would show two particles: the nucleus and the single electron.

But if we were to develop the electron microscope into a moving picture camera of some sort, would it then be possible to follow the movement of the electron in its orbit around the nucleus, and to determine this orbit? Here we are faced with a fundamental difficulty, which makes it quite clearly evident that with this atom model we have reached the limits of visualization. For as soon as we have taken the first exposure on our film, we find that we no longer are in the position to take a second picture of the same atom. This is due to the fact that we will never again find it in an undisturbed condition. The atom has been disrupted by the very electrons which enabled us to take the first picture. The reason is that the first impact of the electrons used to take the first photograph tore the very electron of the atom itself out of its intra-atomic bond, so that the atom appearing on the second picture would never, in any circumstances, possibly be the same, unchanged atom. At best, we would discover the electron somewhere far outside, far away from the nucleus.

It is therefore, evidently, fundamentally impossible to observe the orbit of an electron within the atom. But this impossibility is not due to any shortcoming (still remediable) of that postulated ideal microscope, assumed to be as perfect as natural laws would allow it to be, but rather a consequence of those very laws. By the operation of these self-same natural laws, extremely violent methods—as exemplified by the use of an electron microscope, where the electrons are accelerated in their orbits by a very high voltage—are our only means of getting a sharp, well focused picture of an atom which is in conformity with natural laws.

At this point, it can no longer astonish us to discover that we have reached the limits of visualization, and that the concept of electrons circling a nucleus cannot be taken literally, like the concept of a water molecule consisting of two hydrogen atoms and one ogygen atom, arranged in a triangular pattern.

The limitations imposed here on visual representation can be formulated with greater accuracy with the aid of a relationship, called the *uncertainty principle*, which is based on the quantum theory. It can be expressed in the simplest form as follows: One can never know simultaneously with perfect accuracy both of those two important factors which determine the movement of one of these smallest of particles—its position and its velocity. It is impossible to determine accurately *both* the position and the direction and speed of a particle *at the same instant*. If we determine experimentally its exact position at any given moment, its movement is disturbed to such a degree by that very experiment that we shall then be unable to find it at all. And conversely, if we are able to measure exactly the velocity of a particle, the picture of its position becomes totally blurred.

But by this statement I have anticipated the climax of a development which I must now attempt to describe from its very beginning.

Quite apart from the difficulties involved in obtaining a visual representation, the atom model has certain other properties which do not seem to agree at all with actual experience. For, according to all our previous knowledge, an electron revolving around an atomic nucleus should not be able to revolve in a circular or elliptical orbit for any length of time. In the first place, the electron carries an electric charge, and secondly, it vibrates in its orbit around the nucleus. As the vibrations of electrons in a radio aerial produce an electric wave, the electron vibrating in the atom must emit a wave of radiation, which in this case we should observe as ultraviolet light. But this would mean an emission of energy at the expense of the planetary electron, and the result would be that after a certain length of time, the electron would have to fall into the nucleus and come to rest there. This picture is totally different from the view presented by the atom model with its electrons circling in their orbits undisturbed. And even if it were possible in some way or other to get over this question of radiation, Rutherford's atom model would still fail entirely to account for the stability and definiteness of the chemical properties of the atom.

The problem of co-ordinating atomic stability with Rutherford's atom model was solved by Niels Bohr in 1913. Bohr's views were based on a theory of Max Planck. In 1900, Planck had formulated—at first purely empirically—his radiation law, which interpreted the thermal radiation of a 'black body' (a body which absorbs all the radiation incident on it, and consequently—according to a law formulated by Kirchhoff—is also the most powerful emittor of radiation) in conformity with actual experience. In his subsequent attempt to give this law a physical interpretation, Planck encountered a very peculiar discontinuity in natural processes. He found that he was able to establish his radiation law solely on the foundation of the remarkable hypothesis that the smallest radiating particles, the atoms, could not assume the continuous sequence of all possible energy values of their vibrations (as they would have been expected to, according to all previous knowledge) but only a series of certain definite, specific energy values. It even appeared —and later research proved it, too—that the emitted radiation manifested this quality of discontinuity, and that light, regarded up to then as a wave process, must also consist of discrete quanta of energy. Planck realized that light of the frequency ν must be emitted as well as absorbed in discrete quanta of energy, of a magnitude proportional to the frequency and he postulated that the energy of these quanta was equal to $h\nu$. This constant, h, is called *Planck's constant* (see Table I), and has been the keynote of the entire development of physics for several decades.

As a result of this theory of energy quanta of light, a very strange situation ensued. On the one hand, it was recognized clearly that certain optical phenomena—e.g. those of interference—could only be understood by regarding light as a wave process. On the other hand, the concept of light quanta appeared to be no less necessary in order to explain other phenomena. But according to the latter view, light has no longer the characteristics of waves, propagated in space in all directions, but is a swarm of particles traversing space in straight lines. Therefore, it appears to be impossible to do without two completely different and fundamentally contradictory views of light. We can thus speak of a *wave-particle duality*—of a wave aspect and a corpuscular or particle aspect of light.

Bohr started out from the above-mentioned hypothesis of a discontinuity in atomic phenomena, and he advanced the theory that atoms can exist for an appreciable length of time in certain specific states only, which are characterized by specific electron orbits—in other words, by definite energy levels of their planetary electrons—and that when they are in these stationary states, they do not radiate. This theory explained the stability of atoms, but there was still a long way to go before the natural laws governing atomic structure could become known.

During the first decade of the existence of Bohr's theory, the chemical properties of the elements were finally explained on the ground of the quantum theory. A further important step was successfully taken by the Frenchman Louis de Broglie, who realized, in 1924, that this strange duality of character which at times gave light the aspect of a wave, at times that of a swarm of particles, was a property not only of light, but of matter, too. This discovery finally led to the formulation of wave or quantum mechanics, which in a sense brought the theory of the extranuclear structure of the atom to completion.

Let us now try to elucidate this so-called duality of matter by two photographs. To begin with, let us observe once again the tracks of alpha particles in Figure 3, where their particle character is most clearly evident. After a study of these tracks, one can no longer have the slightest doubt that very small particles have actually flown through space and been deflected at some point here.

On the other hand, we have the evidence of experiments which indicate, with the same degree of certainty, that alpha rays are not particles but waves propagated from the source of radiation. I will not demonstrate this by the alpha rays, but by the beta rays, the particle character of which has been proved by numerous experiments no less convincingly than that of alpha rays. If beta rays are allowed to penetrate thin layers of matter, they show exactly the same kind of interference phenomena as may be observed in the case of x-rays under the same conditions—and we are accustomed to regard x-rays as typical wave radiation. The interference phenomenon consists in the fact that a central ray which penetrates the layer of matter, is surrounded by interference rings, which can be interpreted only as a superimposition of deflected waves. For comparison, we show here two photographs, one (Figure 5) taken by x-rays, and another (Figure 6) taken by beta rays. The latter one is one of the first photographs of this kind ever taken. But for the difference in their sharpness these two photographs would be indistinguishable.

As we see, it is impossible to entertain any doubt whatsoever as to the remarkable dual character of the electron. On the one hand, electrons may be regarded quite legitimately as particles, and we can be absolutely certain that a snapshot of an atom will be a picture similar to that shown in Figure 1. But on the other hand, they appear as waves, too, and this wave nature can also be used to obtain a picture of the atom—

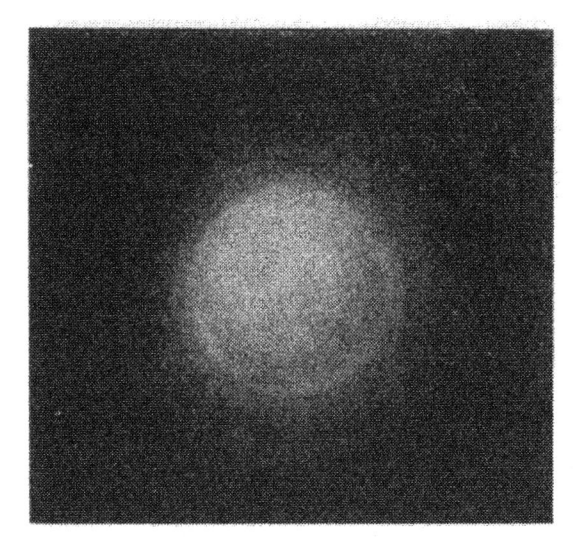

Figure 5.—Interference rings of Röntgen rays.

although in this case, the picture will look considerably different.

Modern physics utilizes both the wave aspect and the particle aspect, and treats both as of equal right. For we know when to use one and when the other, and we know also that neither can stand by itself, without the other.

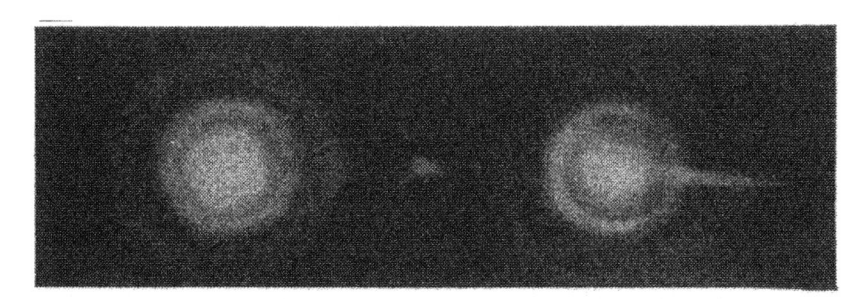

Figure 6.—Interference rings of Beta rays.

Thus we see that it is very difficult to obtain a visual concept of an atom and at this point, we have evidently reached the limit of the possibilities of visualization. As we have seen, the most that can be observed of an atom, is the result of one single snapshot. But such a snapshot never shows the orbit of the electron, but two specific points only—the nucleus and the electron in its momentary position. If a great number of such photographs are taken, in succession, of different atoms of the same kind, the electron will be found in various locations, at a greater or lesser distance from the nucleus, more frequently in one spot, less frequently in another. In this way, we eventually obtain a general, overall picture of the probability of finding the electron at one point or another in the vicinity of the nucleus, a *probability value* for the distribution of electrons. But all these photographic records as a group—being collected in a single instant, as it were—can be regarded also as a picture of the average distribution of the density of electricity in the vicinity of the nucleus. Such a density distribution is comparable, in some degree, to a wave phenomenon—a stationary wave. If we conceive of stationary waves of electrically negatively charged matter, these, too, will correspond to a certain specific density distribution of this matter. Actually, the de Broglie matter waves can be interpreted so that the square of the amplitude of the waves at a certain specific point indicates the density of matter at that spot. But we can just as well say that it indicates the possibility that in a snapshot the electron will be discovered at that very point. These stationary waves in the immediate vicinity of the nucleus were investigated by Schrödinger; they represent, in fact, the object of his wave mechanics. These stationary waves also form the foundation for verifying a stationary distribution of electricity in the vicinity of the nucleus, and consequently, the existence of stationary states of the atom, in which no radiation occurs, becomes understandable. Thus one of the difficulties in connection with the atom model is eliminated.

The individual stationary waves which can possibly occur in an atom, do not form a continuous series, any more than do the individual characteristic vibrations of the strings of a musical instrument. A string can vibrate in its fundamental tone, in which case it has no nodes. On the other hand, it can vibrate also in an overtone, and in that case it has one or more nodes. Similarly, an atom can 'vibrate' in its ground state, in which case it has no 'nodes'—meaning, no levels on which the density of matter disappears. But it can vibrate also in an overtone—in an 'excited' vibration, as it is called—and then there are several nodal levels of zero density. These various stationary vibrations correspond to the various stationary states which an atom is capable of assuming.

Let us elucidate the conditions described above by the example of the simplest of atoms, the hydrogen atom, with the aid of illustrations. The ground state of a hydrogen atom (denoted by the symbol $1s$) was described by Bohr in 1913 as a circular orbit of the electron around the nucleus (Figure 4). In the particle aspect, this is a clear picture. According to this view, the electron has a spin moment about the nucleus. To-day we know that actually it has none. Therefore, to-day we say rather—still within the framework of the particle aspect—that the electron shuttles on a straight line about the nucleus. Thus, we imagine the atoms as shown in Figure 7 (*a*).

But on the other hand, we can conceive of the state of electronic matter in a hydrogen atom also as a wave process (Figure 7 (*b*)). If we should take thousands of snapshots of the atom in its ground state and then develop them superimposed on each other, we would get a density distribution, or probability distribution, such as is shown in Figure 7 (*b*). It is computed from Schrödinger's wave mechanics.

But there exist also 'overtones'—excited states—through which an atom can pass as a result of the impact of a foreign electron. In such an excited state, the atom becomes capable

MOLECULES AND ATOMS 37

of radiation, of emitting a *light photon* (quantum of energy of light). This happens because the atom passes from its excited state back to its ground state, or into another excited state of smaller energy. Figure 7 (*a*) and (*b*) show such excited states, marked by the symbols 2*p* and 2*s*. In the wave aspect of matter, the excited state closest to the ground state is represented by a

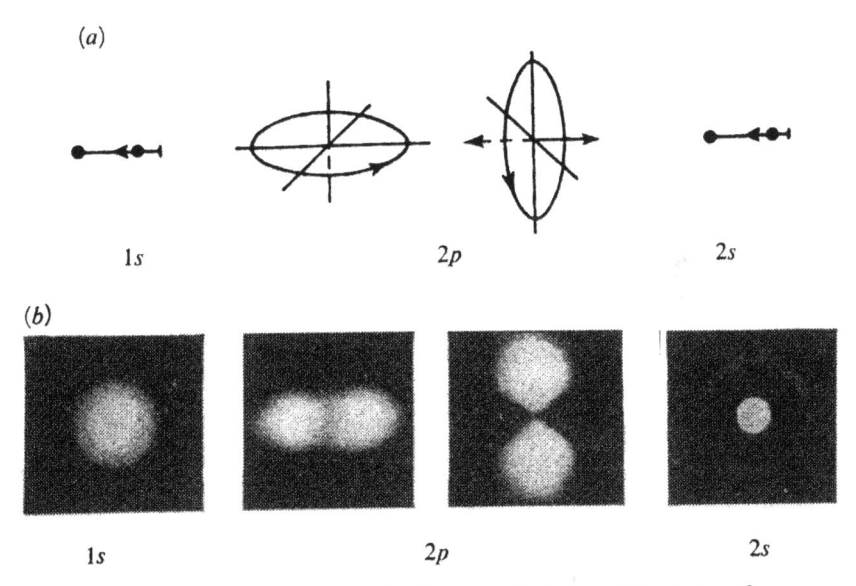

(*a*)

1*s*   2*p*   2*s*

(*b*)

1*s*   2*p*   2*s*

Figures 7—Hydrogen atom in the ground state and in states of excitation.

stationary wave with a nodal plane perpendicular to the plane of our drawing. But this density distribution is, again, just a model, a mere aid to our visual perception, and acquires a concrete significance solely by virtue of thousands of snapshots. If we want to depict the same stationary states under the particle aspect of matter, we obtain the picture of a circular orbit, as in Bohr's original theory. But this orbit can assume many different spatial positions, and the superposition of all these possible states results in a probability or density distribution, in which the same nodal plane appears which is present in the wave aspect.

For excited states of higher energy ($2s$), a high degree of density is obtained in the centre, with a sparsely occupied ring on the outside. In such cases, there is always a certain probability that the electron may suddenly be encountered outside, in the vicinity of the ring.

We shall not go into any further details here. My sole purpose has been to give you an approximate idea of the different concepts employed in physics these days in order to illustrate the structure of the atom, in some degree at least. The reason why no such visual representation can ever manage to cover simultaneously every characteristic of atomic structure, has been discussed already.

III. THE PERIODIC SYSTEM OF ELEMENTS

We come now, finally, to the question of the relationship between the chemical properties of the elements and the structure of the extranuclear parts of their atoms, the number of planetary electrons, and thus, ultimately, the nuclear charges of these atoms. We are indebted to Bohr's theory for our understanding of this interrelationship, and we can obtain the most satisfactory general idea of it by arranging the elements according to the magnitude of their respective nuclear charges, in other words, according to their atomic numbers, which are always identical with the number of elementary quanta of positive electricity on their respective nuclei. (Table II, at the end of this book.) Thus, we start with hydrogen (1), continue with helium (2), and proceed in this manner until we reach curium (96). It has been known to chemists for about a hundred years that the chemical properties of the elements repeat themselves when they are arranged in the order of their atomic numbers. If we break off the series at the end of each such period and start a new line, we get the well known *periodic system of elements* of Newlands, Mendelejeff and Mayer (Table III, at the end of this book). According to Bohr, this periodicity of the elements can be explained on the ground of atomic theory as follows:

According to a principle formulated by Pauli, one orbit cannot be occupied by more than one electron. There is room for one electron only in any one orbit (or in any one stationary vibration with a specific quantum number). This principle may be cited here without proof. So, if an atom has several electrons, the electrons after the first one will be located in ever farther outlying orbits. In considering such problems, it is always advisable to refrain from imagining a completed atom and to proceed, instead, outward from the bare nucleus, and to imagine that one electron is added after another, outwards from the centre, until finally the number of electrons characteristic of the element is completed.

If we continue with the successive addition of individual electrons according to this structural principle, we find that after the incorporation of a specific number of electrons, these electrons constitute, in an important sense, a complete system, the addition of further electrons begins, as it were, a new structure at a considerably greater distance from the nucleus. The extranuclear electron structure of the atom is said to be built up of a number of individual *shells*. Those chemical elements in which the extranuclear atomic structure is, itself, completed by the completion of such a shell, hold a special significance. They are the *inert gases*. The first one of these, helium, may be imagined, under the particle aspect of matter, as a nucleus with two electrons revolving about it, at approximately equal distances (Figure 8 (*a*)). Thus the first shell is completed by two electrons already. The next element, lithium, has one electron more, and this third electron moves in a much wider orbit, as a lone electron in a new shell (Figure 8 (*b*)). It is evident that this atom can yield up an electron very easily, and consequently, occurs very frequently as a positive ion. This is the explanation of the electro-positive character of lithium, the most important characteristic of the chemical behaviour of this element.

And so on further. After a certain number of elements, we always find a completed shell, as in helium. The chemical significance of this fact is that these elements—the inert gases, helium, neon, argon, krypton, xenon—do not react chemically at all. They always represent the end of a period, as shown in Table III at the end of this book.

As we have stated before, the first period consists of only two elements: hydrogen and helium. The second period con-

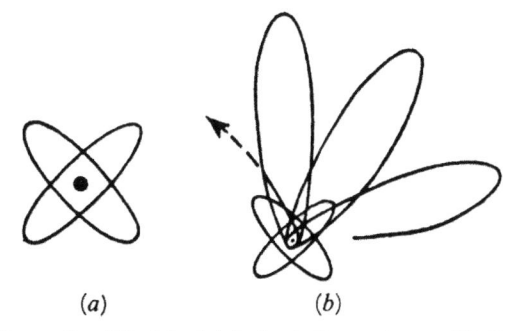

(a)   (b)

Figure 8.—Model of (a) the helium atom; (b) the lithium atom.

tains eight elements: lithium, beryllium, boron, carbon, nitrogen, oxygen, fluorine and neon. Since the element fluorine contains seven electrons in its outermost shell, one more is needed to complete this shell. This fact gives a clue to the chemical properties of fluorine: The fluorine atom always tends to complete this outermost shell by taking up an eighth electron, and thus to have an electro-negative character, and to occur in solutions negatively charged, as a rule. Obviously, those elements which occur at the beginning of a period (hydrogen and the alkali metals), and which therefore will readily give up an electron, combine particularly readily with

MOLECULES AND ATOMS

one of the penultimate elements of the periods, the halogens, which will readily take up an electron. Examples of these combinations are hydrogen fluoride (HF) and sodium chloride, ordinary table salt (NaCl).

The third period, which begins with sodium and ends with argon, again consists of eight elements. Thereafter, the periodic system becomes a little more complicated. Both the fourth period (from potassium to krypton) and the fifth one (from rubidium to xenon) contain eighteen elements each, and the sixth period (from caesium to the inert gas radon) consists of thirty-two elements. The final period is evidently an incomplete one, which for the time being ends with curium. The numbers of elements contained in the various periods—2, 8, 18, 32, i.e. $2 \times 1^2, 2 \times 2^2, 2 \times 3^2, 2 \times 4^2$—obviously follow a simple mathematical sequence. The latter can be explained by the quantum numbers of the individual states of vibration, already mentioned. But we cannot discuss the details of this explanation here.

We have now obtained a general, although merely superficial, survey of our current knowledge of the extranuclear atomic structure. This knowledge enables the physicist to understand the chemical properties of the individual elements, in their general outlines at least. Theoretically, with the aid of quantum mechanics, it would be possible to calculate, quantitatively, every chemical magnitude, such as emission of heat, affinities, etc. But the mathematical difficulties are, as a rule, so great that such calculations have only been actually carried out in some of the simplest of cases.

This brings us to the conclusion of our discussions of the extranuclear structure of the atom. We shall now turn our attention to our principal topic, the nucleus of the atom.

3. RADIOACTIVITY AND THE BUILDING BLOCKS OF THE NUCLEUS

I. RADIOACTIVITY

When investigating the internal physical properties of any system, we must endeavour, on the one hand, to discover its effects on the outside world and, on the other hand, try to approach it in some way that will show us how it behaves during this process. In certain cases, it becomes necessary to dissect it into its component parts, by means of interference from outside. This principle applies to the atomic nucleus, too. Thus there arises the question whether there are any nuclear phenomena which furnish us with the desired knowledge of their internal structure without our having to resort to such interference. The radioactivity (a phenomenon already mentioned) of certain heavy elements is actually such a phenomenon. For this reason, let us discuss it first.

We have already discussed the fact that in radioactivity three distinct types of radiation can be observed: alpha, beta and gamma radiation. A few years after the first observation of radioactivity, Rutherford and Soddy made the discovery, of decisive importance for the development of atomic theory, that the emission of alpha and beta rays is associated with a transmutation of the chemical elements. An atom which has emitted either of these radiations is no longer an atom of the same original element.

This discovery was of paramount importance for atomic theory. The old notion of atoms had now to be discarded; the atoms of chemistry were, evidently, no longer the ultimate, indivisible building blocks of matter. To be sure, it was still not possible to change one element into another by chemical means, yet the existence of a natural process producing this very result appeared to be a certainty. The hopes and aspirations of the alchemists of past ages were thus given new life. For if in certain cases, nature itself effected a transmutation of the elements, it was bound to be possible to perform this process artificially, once the proper tools were discovered. It was bound to be possible, theoretically at least, to turn mercury into gold!

In view of our present knowledge, in the first place that the alpha and beta particles carry electric charges, and secondly, that the chemical properties of the atom depend on the number of elementary charges of electricity on its nucleus, this discovery of Rutherford and Soddy is easily understandable to-day. The alpha and beta particles originate in the nucleus of the atom, and not in its outer structure. The alpha particles are helium nuclei, identified as such by their mass and charge. Their mass is 4 atomic mass units, and they carry two elementary charges of electricity. We express this shortly by saying that their *mass number* is 4, and their atomic number is 2. Accordingly, the helium atom is represented by the symbol $_2He^4$, where the superscript denotes the mass number and the subscript the atomic number. This kind of symbolism is applied to the atoms of all the elements. When an alpha particle is ejected by an atomic nucleus, it carries with it not only its own mass, but also its own charge. The nucleus loses both this mass and charge; its atomic number (the number of elementary quanta of electricity present on it) is reduced by 2, whereas its mass number is reduced by 4. On the other hand, a beta particle is a negative electron. Its mass number is approximately 0, and its atomic number is -1. Accordingly, if e stands for 'electron' in general, the symbol of the negative electron may be written as $_{-1}e^0$. Thus, when an electron is emitted, the mass of the nucleus remains practically unchanged, whereas its positive charge increases by 1, due to the loss of one elementary quantum of negative electricity. Hence, both the emission of an alpha particle and the emission of a beta particle result in a change in the atomic number. Since the chemical properties of the elements are determined by the nuclear charge number, the atomic number, it is evident that both alpha and beta radiation must result in a transmutation of elements.

Let us now examine these facts more closely, using radium as our example. The mass number of radium is 226, its atomic number is 88, and consequently, its symbol is $_{88}Ra^{226}$. The radium atom contains 88 extranuclear electrons; since 86 electrons form completed shells, the chemical properties of radium are determined largely by the two electrons which revolve around the nucleus in the outermost, incomplete shell. Therefore, the chemical properties of radium are similar to those of one of the alkaline earth metals, such as barium or strontium. The radium nucleus is an emitter of alpha rays, and as a result of this radiation, its mass is reduced to 222, and its nuclear charge, its atomic number, to 86. A new element—likewise radioactive—is formed: the inert gas radon, the symbol of which is $_{86}Rn^{222}$. Due to its nuclear charge number, the radon atom contains only 86 electrons, which are arranged in completed shells. This atom is therefore chemically inactive—this element is an inert gas. The process of the emission of an alpha particle by the radium atom and the formation of a radon atom are indicated by the following formula:

$$_{88}Ra^{226} \rightarrow {}_{86}Rn^{222} + {}_{2}He^{4}$$

The symbol to the left of the arrow is that of the radiating atom; the symbols following the arrow show what has become of this atom as a result of the emission of the alpha particle. In such a formula, the sum total of the superscripts must be equal on both sides of the arrow; in this particular case, $226 = 222 + 4$. This follows from the law of the conservation of mass. The same rule applies to the subscripts; in this case, $88 = 86 + 2$. This follows from the law of the conservation of electric charges. Analogous formulae are used for processes in which beta radiation occurs.

All—or nearly all—of the alpha particles emitted by a homogeneous radioactive substance have exactly the same range. This is evident in the cloud chamber photograph shown in Figure 3 (a). It shows the successive decay of two radioactive elements; therefore, we see two groups of alpha rays. The range in air of the alpha particles of different radioactive substances is between 1 and 9 centimetres, and the range is evidently dependent on the energy with which the particles emerge from the nucleus. The greater the energy, the longer is the range.

The various radioactive substances show great differences in their respective speeds of transmutation. Some of these substances are very short-lived, whilst others last very long and show no noticeable lessening in radioactivity over long periods of time. Obviously, for the atoms of every radioactive substance there exists a probability, capable of being expressed numerically, of their radioactive decay. The reciprocal of this probability is the *average life* of the substance. The decay probability, and hence also the average life, is independent of the number of atoms already decayed. This means that the same percentage of the number of atoms still intact will decay per unit time. This law is expressed by the following equation:

$$dN = - \lambda N \, dt$$

with the following solution for N:

$$N = N_0 e^{-\lambda t}$$

where N_0 is the number of the intact atoms present at the time $t = 0$, N is their number at the time t, e is the base of natural logarithms, and λ is the decay probability, and hence $1/\lambda$ is the average life. Instead of the latter, the *half-life* period, T (that period of time during which exactly one-half of the original number of atoms decays) is frequently used. The half-life is slightly less than the average life; it differs from the latter by the factor log nat 2, the natural logarithm of 2. (If we write $t = \frac{1}{\lambda} \log \text{nat } 2$, then $N = N_0 \times e^{-\log \text{nat } 2} = \tfrac{1}{2} N_0$). This law applies to both alpha and beta radiation.

Thus the radioactive properties of a homogeneous substance are determined principally by two factors: the nature of the emitted particles and the average life or half-life of the substance.

The gamma rays play a somewhat different part. We must point out, first of all, that in natural radioactivity gamma rays do not appear alone, but only in combination with one of the other two types of radiation. Gamma rays have an even greater penetrating power than either the x-rays (to which they are essentially analogous in other respects) or the alpha and beta rays. To give an approximate idea of their power of penetration, let it be said that while an alpha ray is absorbed by a single sheet of paper, 100 such sheets would be required to absorb a beta ray, and several thick volumes to absorb the gamma radiation. As already mentioned, gamma rays cannot be deflected, nor—unlike the alpha and beta rays—can they be made visible in the cloud chamber. For although the gamma rays too cause ionization in air, this is not a direct, primary process, but an indirect ionization through the agency of the electrons which the gamma rays dislodge. In the cloud chamber, we can see only the tracks of the secondary particles generated by them; the tracks of the gamma rays themselves are not visible. These two facts are quite compatible with each other, since both can be attributed to the absence of an electric charge in gamma rays.

Actually, gamma rays differ from x-rays, and even from visible light, too, merely in that the wavelength of the gamma rays is much shorter. They are a species of electro-magnetic waves, among which the radio waves have the longest wavelength. At any rate, we know that the already mentioned wave-particle duality applies to all these types of radiation. Hence, while we have just referred to gamma rays as an electromagnetic wave issuing forth from the atomic nucleus, we may just as well look at them under their particle aspect and speak of particles, extremely energetic photons, ejected from the nucleus and speeding through space with the velocity of light.

The fact that under certain conditions atomic nuclei can emit gamma rays, is quite understandable. We already know that the extranuclear structure of the atom can emit light when—for example—the atom is excited due to a gaseous discharge. Atoms emit x-rays also when particles are dislodged from their inner electron shells by extremely fast electrons. This follows clearly from the fact that the extranuclear structure of the atom constitutes an electrical system, and when any such system is disturbed, electromagnetic waves are emitted. But the atomic nucleus, too, is an electrical system, as is shown by its charge, and so it can be reasonably expected that in conjunction with certain internal processes within the nucleus, the nucleus, too, will emit electromagnetic waves—gamma radiation.

In natural radioactivity, beta particles make an appearance always as carriers of a negative charge only, in other words, only as *negative* electrons. Let us jump ahead for a moment and mention that in the radiation of artificially produced radioactive substances *positively* charged electrons are observed also; they are particles having the same mass as the negative electrons, but each of them carries an elementary quantum of positive electricity (*Anderson*). To-day, these particles are known as *positrons*, and electrons are usually regarded merely as a negative variety of them. The question now arises, why these positrons were not observed a long time ago, and why they do not occur in the extranuclear structure of the atom. The answer to this question is based on experiments which have established that positrons are very short-lived. As soon as a positron approaches an electron—usually after a very brief interval of time—it combines with the latter, to form an electrically neutral structure; the product of this

48 NUCLEAR PHYSICS

union is one or two gamma-ray photons, i.e. photons of an extremely short wavelength. These constitute what is known as *annihilation radiation*. Both the existence of positrons and of annihilation radiation were predicted by Dirac, and subsequently confirmed by experiments. The annihilation radiation has its counterpart, too; a photon entering the powerful field

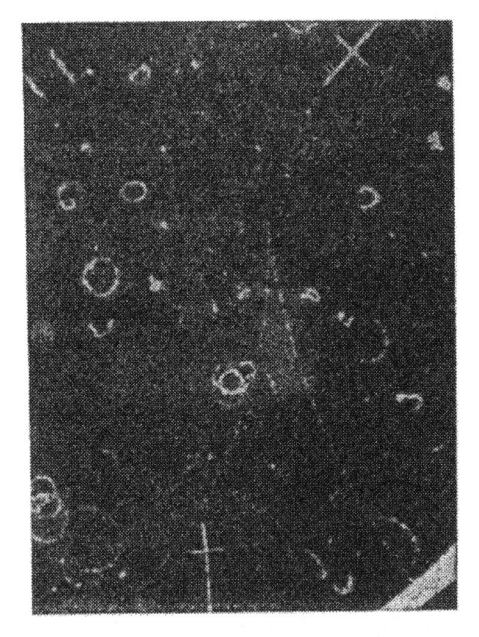

Figure 9.—Pair production in the cloud chamber

in the immediate vicinity of an atomic nucleus, can change into an electron and a positron. This *formation of a pair of charged particles* can be observed in the cloud chamber, and is shown in the photograph reproduced here as Figure 9. In the cloud chamber, a very strong magnetic field deflects the electrons to one side, and the positrons to the other, so that they describe circular tracks (Figure 9). In the upper half of our photograph, such a pair formation has just taken place, and the two tracks of the electron and positron thus formed can be clearly seen. Since this photograph is considerably enlarged, the individual cloud droplets along the tracks can also be recognized. Another electron, the track of which is slightly blurred, is visible, too. The rest of the droplets are due partly to impurities.

However, this phenomenon of pair formation must not be regarded as indicating that a photon is actually composed of an electron and a positron. A photon is a true elementary particle, in the strictest sense of the word. But it is capable of changing, when it enters into interaction with other particles or with powerful fields. Generally speaking, the concept of 'elementary particles' in modern physics has undergone a change; these elementary particles may be described as the 'ultimate, indivisible building blocks of matter' in a very limited sense only. For it has been proved that these elementary particles can change into each other practically without restriction, so long as it is compatible with the laws of conservation of mass, energy, etc. But just for this very reason, it is meaningless to describe any of them as being composed of some of the others.

Beta particles have a far longer range than alpha rays. The explanation is not that the former have more energy, but principally that due to their smaller charge and greater velocity, they have a considerably smaller capacity to produce ionization, and therefore they lose their energy much more slowly along their path.

But there is another very characteristic difference between alpha and beta rays. All alpha rays of a homogeneous radioactive substance have exactly the same range, and therefore, the same energy. This is to be expected, for as the energy released in any chemical reaction is determined both by the initial and end states of the system, so must also the energy released by radioactive decay—in other words, essentially, the energy of the alpha particle—depend solely upon the initial and end states of the atomic nucleus. Generally speaking, all nuclei of the same kind have the same energy, but in the case of beta rays the situation is different from that of the alpha rays. Any homogeneous radioactive substance will emit beta particles with all possible velocities, from an upper limiting velocity down to quite low speeds. It seems that the energy corresponding to this upper limit is identical with the difference between the energies of the atom in its initial and end states. The occurrence of slower particles would, however, contradict the energy principle, unless the energy missing from the individual beta particles were removed from the nucleus in some other way. This brings us to the theory of Pauli, that with every beta particle another particle leaves the nucleus, carrying the difference in energy. The sum total of the energy is always constant, and this total energy is shared by the beta particle and this new particle, in accordance with definite statistical laws.

This new particle must be devoid of all electric charge, for otherwise it would be impossible to explain the fact—confirmed experimentally in every instance—that when a beta radiation takes place, the nuclear charge increases by one unit. The absence of an electric charge on this new particle is indicated also by the fact that these particles cannot be observed in the cloud chamber. Since this new particle is electrically neutral, and its mass is certainly very small, it is called the *neutrino*. According to the evidence of every experiment hitherto performed, the mass of the neutrino is smaller than the mass of an electron. But whether it is actually 0, like the rest mass of the photon, cannot be stated with any certainty as yet.

II. ARTIFICIAL NUCLEAR TRANSMUTATIONS

We have thus described those processes in which the atoms themselves provide us with clues to the properties of their nuclei; now we shall proceed immediately to consider those experiments aimed at gaining more exact and detailed information concerning the nucleus, by means of interference from outside. Again, it was Rutherford who took the first step in this direction, too. He discovered the proper tool for the artificial transmutation of one atom into another—bombardment of atoms by alpha particles. In 1919, he achieved by this method the transmutation of an element: he turned nitrogen into oxygen. But let it not be assumed that this method offered the means of transmuting ponderable quantities. The transmutation affected a very small number of atoms only; but, of course, this does not detract from the fundamental significance of this discovery.

Rutherford found that a certain type of radiation, consisting of positively charged hydrogen atoms—nuclei of hydrogen—was emitted when nitrogen atoms were bombarded by alpha rays. The nucleus of a hydrogen atom is an elementary particle, and as we shall see later, it is—for this very reason—one of the most important fundamental building blocks of matter. Consequently, it has been named *proton*. When nitrogen atoms are bombarded by alpha particles—in other words, by helium nuclei—a proton is emitted occasionally by the nitrogen nucleus. The alpha particle remains in the nucleus. The laws of the conservation of mass and energy permit us to calculate what happens to the nucleus in such a case. The nitrogen nucleus has the mass number 14 and the nuclear charge number (atomic number) 7; its symbol, therefore, is $_7N^{14}$. The mass number and charge number of the alpha particle are, respectively, 4 and 2, and its symbol is $_2He^4$, as already mentioned; the symbol of the proton is $_1H^1$. As regards mass, the nitrogen nucleus is changed by the absorption of an alpha particle and the loss of a proton, as follows:

$$14 + 4 - 1 = 17$$

But as regards its charge, the following equation applies:

$$7 + 2 - 1 = 8$$

Thus, a nucleus of mass 17 and nuclear charge (atomic number) 8 is formed. This nuclear charge number indicates that it is an oxygen atom, but the mass number 17 does not agree with the mass number of the common oxygen atom, which is 16. The truth is that this is a rare *isotope* of oxygen. We shall discuss such isotopes in due course. This transformation of $_7N^{14}$ into $_8O^{17}$ is the renowned first instance of the artificial transmutation of elements.

Such nuclear reactions can be represented by appropriate formulae. The formula of this particular reaction is:

$$_7N^{14} + {_2He^4} \rightarrow {_8O^{17}} + {_1H^1}$$

Such nuclear transformations can be observed in the cloud chamber (*Blackett*). But since these are very rare phenomena, many thousands of photographs must be taken in order to be lucky enough eventually to see a nuclear transmutation take place.

In Figure 10 we see a large number of alpha particle tracks, going from right to left. But at one point, an individual diagonal track crosses the main path of the alpha particles, travelling upwards and slanting to the right. If we study the photograph more closely, we can detect a second very dense track, which starts out from the point of origin of the former and runs diagonally to the left sloping slightly downwards. This point is where a nuclear transmutation took place, produced by an alpha particle. These two tracks are: first, the path of the proton knocked out of the nucleus, and, second, the path of the transformed nucleus which received a powerful impact in this process.

These experiments indicate that protons are, in all probability, the fundamental building blocks of atomic nuclei, and this is reminiscent of Prout's hypothesis that all atoms are formed from hydrogen.

Now in 1932 a particle of a previously totally unknown kind was discovered, which can be knocked out of atomic

Figure 10.—Transformation of a nitrogen nucleus into an oxygen nucleus.

nuclei by an analogous method. This discovery was the achievement of three scientists, Joliot, Curie and Chadwick, who followed a path first taken by Bothe in Germany. This particle, of practically the same mass as a proton, carries no electric charge, and therefore leaves no visible track in the cloud chamber. It was named *neutron*. The first nuclear reaction in which the emission of a neutron could be observed, was the transmutation of beryllium. Beryllium atoms, mass number 9, atomic number 4 ($_4Be^9$), were bombarded by alpha particles, $_2He^4$; the product was a carbon nucleus of mass number 12 and atomic number 6. The following equations show the masses and charges (mass numbers and atomic numbers) of the particles involved:

$$9 + 4 - 12 = 1 \text{ and } 4 + 2 - 6 = 0$$

As we see, in this process, a particle of mass number 1 and charge (atomic) number 0 is ejected—a neutron. We designate this particle by the symbol $_0n^1$, and we can express this *nuclear reaction* (this is the name given to such processes) as well as those previously discussed, by formulae. We take into consideration also that in addition to the neutron a gamma-ray photon (its symbol is γ) is frequently emitted, too. The formula of the nuclear reaction just mentioned reads, therefore:

$$_4Be^9 + _2He^4 \rightarrow {_6C^{12}} + _0n^1 + \gamma$$

In such nuclear reactions it frequently happens that unstable atoms are produced; radioactive atoms, not occurring in nature, which change after a certain length of time into some stable atomic species. In the cases hitherto known, this process occurs with emission of electrons or positrons only. This fact completes our survey of the nature of the particles which issue forth from atomic nuclei, whether by spontaneous emission or due to outside interference.

III. THE BUILDING BLOCKS OF ATOMIC NUCLEI

With the knowledge which we have gained we can now proceed to consider the question, which of the elementary particles can be regarded as the ultimate building blocks of atomic nuclei. Let us, first of all, enumerate these particles once again? The *proton* and the *neutron*, the *electron* and the *positron*, the *neutrino*, and the *gamma-ray photon*. There are, furthermore, the *alpha particles*. But in view of the mass and charge numbers of the latter, we may assume that they are not elementary particles at all, but composite structures. However, our list of elementary particles is still incomplete. In the first place, it does not include the *antineutrino*, which is the opposite number of the neutrino, as the positron is the opposite number of the electron. While it has the same near-zero mass and zero charge, it differs from the neutrino as regards a property which is present in all elementary particles, and which we have not mentioned as yet: Its *spin* (or as it is called technically, its *angular momentum*) is opposite in direction for a given direction of the magnetic moment. Many elementary particles behave mechanically like small spinning tops. But their angular momenta can have only certain definite values, which can be accounted for by quantum mechanics. In the cases of the elementary particles with which we are concerned here, this value is, generally, $ℏ/2$ or $ℏ$; this $ℏ$ is an abbreviated symbol for $h/2\pi$, and h is Planck's constant. As a general rule, the effect of this spin moment is that the particles possess a magnetic moment, in other words, they act like small magnets. In the case of heavier particles, this magnetic moment is measured in terms of a unit called *nuclear magneton* (*n.m.*), while for lighter particles a larger unit, called *Bohr magneton* (*B.m.*) is used. (See Table I.)

There is still another elementary particle we must mention; it is called *meson*, for its mass is between that of an electron and that of a proton (the Greek word 'meson' means 'the middle one'). Mesons can be observed in cosmic radiation, and before we go into further details about their properties, we must first say a few words about the nature of cosmic radiation.

As a result of the studies of Hess and Kohlhörster, we have known for almost forty years that a faint, continuous, extremely penetrating radiation reaches the earth from outer space, and in the atmospheric envelope of the earth it releases every possible sort of secondary radiation, akin in character to radioactive radiation. However, it is only since 1947 that we have known anything at all about the cause of this remarkable phenomenon. This knowledge is the fruit of the work of Forbush and Ehmert, who were able to prove that when certain eruptions take place on the surface of the sun, the cosmic radiation incident on the earth suddenly gains in intensity. Thus, cosmic radiation is probably produced by big, periodically changing electromagnetic fields on the surfaces of the stars, and (according to Unsöld), on the surfaces of the many red stars in particular, or in intersteller space. These electromagnetic fields can assume very large magnitudes and accelerate charged particles to extremely high velocities (according to Bagge and Biermann, especially when a strong spot activity exists on the stellar surface). The hydrogen nuclei are the ones that are accelerated this way in the first place, for the stars consist mostly of hydrogen, but this effect is imparted also to the nuclei of heavier atoms; they leave the stars and speed through cosmic space as the *primary* cosmic radiation. The particles hit the atmosphere of the earth from outer space, with kinetic energies ranging mostly from 10^9 to 10^{10} electron-volts. But particles of energies up to 10^{16} electron-volts have also been observed. In the earth's atmospheric envelope, particles of such enormous energy produce not only atomic decay but also nuclear transmutations of all sorts, creating various kinds of elementary particles in so doing.

This is how the elementary particle just mentioned, the meson, was discovered (*Anderson*). The importance of this particle for nuclear physics is still uncertain. We know that its mass is approximately two hundred times the mass of an electron, that it carries one elementary charge of electricity, and that there are both positive and negative mesons. Beyond this, not much is known about its properties as yet. In 1947, Powell discovered another elementary particle, the mass of which is about three hundred times the mass of an electron. It is called *heavy meson* or π-particle. The researches of Leprince-Ringuet, Rochester and Butler have shown the probability of the existence of a still heavier elementary particle, of approximately nine hundred times the electronic mass.

Table Ic contains a list of all the above-mentioned elementary particles and their respective properties. By calling them 'elementary particles', we mean that they are not composed of still smaller particles, in contrast to the chemical atoms which obviously can be split up into component parts. But this does not indicate by any means that these elementary particles cannot be transformed. On the contrary, transformability is a characteristic of elementary particles. A photon can change into an electron plus a positron, and conversely, a photon can originate from an electron and a positron. But it would be wrong, or at least not advisable, to say therefore that a photon is a combination of an electron and a positron. For, conversely, a photon can be the product of an electron, for instance, when the electron jumps from one state to another. Moreover, a proton can change into a neutron and a positron, or a neutron into a proton and an electron. But one can hardly say that a proton is made up of a neutron and a positron. All these are true elementary particles, of which convertibility happens to be one of the characteristic properties.

In our survey of these elementary particles, we must look for properties capable of giving us a clue as to which ones are to be regarded as true building blocks of atomic nuclei, and which play a different part. For instance, the fact that a particle can emerge from the nucleus, is still not a proof of its being an ultimate, fundamental component part of the latter. This is evidenced by a simple consideration of the extra-nuclear structure of the atom. It consists of electrons. Yet, occasionally, also other particles emerge from it—namely: photons. But since these make an appearance only when certain changes in state of the extranuclear atomic structure occur, they are not referred to as integral component parts of that structure. Thus we make a distinction between particles which are always present in the extranuclear atomic structure—which we designate as its true component parts—and those which are produced within it, occasionally, due to changes in state, and depart from it subsequently. In the case of the extranuclear atomic structure, the former are the electrons, and the latter are the photons. In a certain sense, though, one might of course say that the photons were already present in that extranuclear structure. For although the space between the electron shells is empty (fast particles can penetrate it without any difficulty), yet it contains something—the electric field, which, to use an analogy, acts like the mortar that binds the building blocks of the extranuclear structure of the atom to the nucleus. In the wave aspect, the emitted light is an electromagnetic wave, the energy of which is drawn from that of this field. Fundamentally, it is merely a question of terminology whether we call this field a kind of substance or a property of space, and in this sense the photons may be said to have already been present as a field in the extranuclear structure. Nevertheless, it is helpful to make a distinction between the true building blocks of the nucleus and the field that holds them together, although this distinction cannot be of any fundamental importance. In any case, with respect to the extranuclear structure of the atom, there is a good reason for referring solely to electrons as its elementary building blocks and crediting the field with the capacity of occasionally producing photons. The following distinction may be made between the two kinds of particles: whenever the extranuclear atomic structure is subjected to outside interference—as, for instance, to bombardment by electrons or photons—the result is either that an extranuclear, planetary electron is detached and hurled out of the atom, or that the extranuclear structure itself remains in an excited state, from which it will revert to its original state through the emission of a photon only. But whereas the electron is ejected in the very moment of the interference, the state in which the photon is formed and emitted does not ensue immediately; a certain length of time must always elapse first. This length of time is very short in absolute duration, but of considerable length relative to the time required for a complete revolution of a planetary electron around the nucleus. We may regard this as a characteristic which distinguishes true elementary particles from secondarily formed ones.

Proceeding from the same viewpoint, let us now examine those elementary particles which merit immediate consideration as building blocks of the nucleus. The nucleus, too, can be subjected to outside interference, in the form of bombardment by alpha particles or other elementary particles. As we have already seen, nuclear reactions may then take place, in which either a proton or a neutron is hurled forth. There are also instances where alpha particles are ejected. But it is certain that the latter are not true elementary particles. The ejection of a proton or neutron—like that of an extranuclear, planetary electron—usually occurs in the moment of interference. But in such a nuclear reaction an unstable, radioactive atom may be produced, which then changes further by a radioactive process. In these cases, only electrons or positrons, as well as neutrinos, are emitted. Like the naturally radioactive substances, these radioactive atoms have a definite decay probability or average life, which varies according to the individual instance. Thus, a longer or shorter period of time must elapse before the electron or positron is ejected, accompanied by a neutrino, as in the case of the emission of photons from the extranuclear atomic structure. On the other hand, there are also cases in which outside interference will result in the emission of a gamma-ray photon. Generally speaking, the time interval is long when measured in terms of nuclear frequencies, but extremely short in absolute duration, and it cannot be measured directly, but only by inference from other data. It occasionally happens, of course, that a gamma-ray photon is ejected in the very moment of the interference,

From the above findings we can conclude that only the protons and the neutrons may be regarded as true building blocks of the nucleus. This conclusion is very close to Prout's old hypothesis. The mass of a neutron differs very little from that of a proton.

The ideas discussed are illustrated in the following table by a comparative survey of the conditions which obtain in the extranuclear structure of the atom, on the one hand, and those present in the nucleus, on the other hand.

	Extranuclear Structure	Nucleus	
Fundamental Building Blocks	Electrons	Neutrons Protons	
Field of Force	Electric Field	Electric Field	Nuclear Field
Particles emitted when changes in state take place	Photons	Photons	Electrons Positrons Neutrinos

This first horizontal row contains the fundamental building blocks, while the second one shows the field acting between them, and in the third one we find the particles emitted in the occasional changes of state.

We have still to inquire into the nature of the force which binds these building blocks of the nucleus together. It would seem logical to assume that this field, like the one within the extranuclear structure, is electrical. It is, however, easy to show that electric forces alone would not be sufficient to provide an explanation of nuclear cohesion, for the simple reason that the strongest electric effects are due to the charges carried by the protons, and these are forces of repulsion. Therefore, a further field of another type must be operative in the nucleus. For lack of a more exact knowledge of the nature of this field, let us first give it a name, and call it *nuclear field*. Of course, in addition to it, there exists also an electric field in the nuclei, since the protons carry electric charges.

In the extranuclear structure of the atom, changes in state are accompanied by the production of particles—photons—out of the energy of the electric field. In considering the nucleus, we shall, analogously, look for particles which come into being, in conjunction with changes in state, out of the energy of its two fields, and which occasionally—in excited states—detach themselves and are emitted. Again, photons alone can correspond to the electric field; in fact, it has been discussed that in nuclear transmutations gamma-ray photons are frequently emitted, which are simply photons of an extremely short wavelength. It is obvious, therefore, that the nuclear field must be correlated with the other particles which are emitted in nuclear transformations after the lapse of a certain period of time only—namely, the electrons, positrons and neutrinos. This analogy between the extranuclear structure of the atom and its nucleus gives us a relatively simple picture of the latter.

A nucleus is built of protons and neutrons. Its building blocks interact firstly, through an electric field which results from the charge carried by the protons, and secondly, through a nuclear field, which in some as yet undisclosed way ensures the internal cohesion of the nucleus. The electric field is responsible for the emission of gamma-ray photons, and the nuclear field is responsible for the emission of electrons, positrons and neutrinos.

Whether the connection between the nuclear field and the particles correlated with it here is as simple as that between the electric field and the photons, is, of course, open to doubt. As we shall see in due course, this relationship is probably a more complex one. But generally speaking, it is permissible to apply the analogy outlined above.

Thus we have obtained a fairly clear picture of the nucleus. We may interpret it as meaning that a sufficiently powerful super-microscope would show nuclei to be such as we have described them to be—built of two kinds of component parts, protons and neutrons. Therefore, every nucleus can be identified extremely simply, by two numbers: the number of its protons, and the number of its neutrons.

The mass of a nucleus is (to be sure, not quite exactly) equal to the combined number of its protons and neutrons; both particles have a mass of approximately 1 mass unit. On the other hand, the nuclear charge is possessed by the protons alone, each of which carries one elementary electric charge.

From this simple picture of the nucleus it follows, in general, automatically that different nuclear mass numbers can be associated with a given nuclear charge (atomic number). In other words, there are different varieties of atomic nuclei for the same chemical element; these different varieties are called the *isotopes* of the element.

The mass number of the nucleus indicates the sum total of the number of its protons and neutrons, whereas the atomic number shows the number of protons only. Hence, the number of the neutrons is simply the difference between the mass number and the atomic number. These two numbers, the essential characteristics of the atom, have already been introduced as a superscript and a subscript to the chemical symbol of the element. For instance, the notation $_7N^{14}$ means that this nitrogen nucleus consists of 7 protons and $14 - 7 = 7$ neutrons.

Let us now examine some of the simpler nuclei more closely. Disregarding the neutron, which is not usually grouped with the elements in the proper sense of the word, the simplest nucleus of an element is the proton itself. The proton is a hydrogen nucleus. Its symbol is $_1H^1$, expressing that it consists of 1 proton and $1 - 1 = 0$ neutron. The chart below shows schematic representations of the simplest nuclei; black dots stand for protons, circles for neutrons. Thus, the hydrogen nucleus $_1H^1$ is represented simply by a black dot. But as Urey discovered in 1932, there exists another nucleus, a heavy isotope of hydrogen, which consists of one proton and one neutron. This isotope occurs in natural hydrogen to the very small extent of 0·02%. This variety of hydrogen is called *deuterium*, and its nucleus a *deuteron*. Since it differs chemically in certain respects from ordinary hydrogen, it is usually designated by the symbol D, or more exactly, $_1D^2$. It may be

Hydrogen			Helium			
$_1H^1$	$_1D^2$	$_1T^3$	$_2He^3$	$_2He^4$	$_2He^5$	$_2He^6$
Proton 99·98%	Deuteron 0·02%	Triton $-\beta$(31 years)	$\sim 10^{-5}$%	100%	$\propto(\sim 10^{-19}$ sec.)	$-\beta$(0·8 sec.)

written also as $_1H^2$. This is the so-called 'heavy hydrogen'. A third variety of hydrogen was discovered later; its nucleus consists of 1 proton and 2 neutrons. It is called *tritium*, and its nucleus is referred to as a *triton*. Its symbol is $_1T^3$ or $_1H^3$. This nucleus is not stable; it is radioactive, with a very long half-life (probably about thirty-one years), and is an emitter of electrons. Therefore, tritium does not occur in nature, but is a product of nuclear transmutation processes.

The next simplest element is helium, the nucleus of which contains two protons. Helium occurs also in the form of several isotopes, each with a different number of neutrons in the nucleus. The lightest helium nucleus consists of 2 protons and 1 neutron, and its symbol, therefore, is $_2He^3$. It became known as the product of the radioactive transmutation of the triton, $_1T^3$, although it occurs also in natural helium, in very minute quantities. (This nucleus and the nuclei of the other helium isotopes, are shown in the above chart.) The next nucleus is that of ordinary helium, consisting of 2 protons and 2 neutrons; its symbol, as we already know, is $_2\text{He}^4$. This is an especially stable structure. There are two other helium nuclei, neither of which is stable. They contain, respectively, 3 and 4 neutrons, and are therefore written $_2\text{He}^5$ and $_2\text{He}^6$. They are not present in natural gaseous helium.

Continuing to add further protons and neutrons, we come to nuclei which are increasingly complex in structure. We could make a chart of all the existing nuclei, recording their atomic numbers—in other words, the number of their protons —designated by the symbol Z as abscissa, and the numbers of their neutrons, designated by the symbol N, as ordinate. It is however more convenient for the printer to adopt a different system in which the abscissae, Z, represent the number of protons and the ordinates, $N-Z$, the amount by which the number of neutrons exceeds that of the protons. This is the method used in our Tables IVa and IVb. The order of the individual elements also corresponds to their sequence in the periodic system. In the tables, the nuclei are also distinguished according to their stability or radiation characteristics. Stable nuclei are represented by black dots. One of these is the nucleus $_2\text{He}^4$, which appears where $Z = 2$ and $N-Z = 0$. Triangles indicate the radioactive nuclei which emit beta rays. When the apex of the triangle points upward, it indicates an emitter of electrons; a triangle with the apex pointing downward indicates an emitter of positrons. The electron emitters always appear in the uppermost row—e.g., $_2\text{He}^6$ or $_3\text{Li}^8$—and the positron emitters mostly in the lowest row, e.g., $_6\text{C}^{11}$. The radioactive nuclei which emit alpha particles, are indicated by small squares. Finally, there are unstable nuclei which change by the capture of an electron from the innermost extranuclear electron shell and thus reduce their nuclear charge number by one unit. In our tables, these nuclei are indicated by small circles. Those nuclei which can emit both electrons and protons, are marked by two triangles superimposed on each other in the shape of a star. Thus, our tables furnish a simple survey of all the existing nuclei, their structures and properties.

As we see, the neutron excess, $N-Z$, is negative in just a few species of atoms only; in all the rest, it is a positive magnitude, and never a very big one. The nuclei of the lighter elements contain, throughout, practically as many neutrons as they contain protons, and the neutron excess becomes more or less considerable in the heavier elements only.

We have now described, in general outline, the structure of individual nuclei, and this discussion brings up a great many further questions: What is holding together a nucleus which consists of protons and neutrons? What is the nature of the forces that bind these particles together? Why do lighter nuclei contain approximately as many protons as neutrons, while the heavier ones show a slight neutron excess—in other words, why does the magnitude of the neutron excess increase with the increase of the number of unit charges on the nucleus? Why is there only a limited number of nuclei? Why are many of these radioactive, and why do they emit those very particles which we observe to be emitted by them? In the following lectures, we shall discuss these problems.

4. THE NORMAL STATES OF ATOMIC NUCLEI

I. THE BINDING ENERGY OF THE NUCLEI

Of the questions mentioned at the conclusion of the previous lecture, let us discuss first the one concerning the forces which are operative between the building blocks of the nucleus and hold these particles together. This question may be formulated as follows: What physical magnitude or what property of the atom determines its stability? One might be inclined to think, at first, that this question is a difficult one, and that in order to answer it, we must be familiar with the entire mechanical system represented by an atomic nucleus, in all its details. However, fortunately, this is not the case. There are a few fundamental laws which enable us to discuss the stability and other general properties of a system even though we may be ignorant of the nature of the forces acting within it, as well as of the details of its structure. These are the well known laws of the conservation of mass, energy, etc., which state that these things can neither be created out of nothing nor be annihilated. Those laws which are of primary significance here, are the laws of the conservation of energy, of electric charge, and of angular momentum.

We shall begin with the law of the conservation of energy. Let us assume that it is possible to employ certain forces so as actually to remove a particle from a nucleus (which latter consists of protons and neutrons). We may make this theoretically possible by the naïve assumption that we can actually seize this particle and transport it to a point far away from the nucleus. Since the particle was originally firmly bound to the nucleus, it is now attracted by it, so that a certain amount of work is required in order to remove it; in other words, energy must be introduced into the system. Now, according to the energy laws, this work, this imported energy, is quite

THE NORMAL STATES OF ATOMIC NUCLEI

independent of the method employed to remove the particle. Therefore it follows that every particle is bound within the nucleus by a definite quantity of energy, and this quantity of energy can be calculated if it is somehow possible to ascertain the quantity of energy of the system before and after the removal of the particle. Let us now proceed to define the concept of the *binding energy* of a nucleus. This term stands for that change in nuclear energy which takes place when the constituent parts of the nucleus—originally far away from each other—are combined to form that nucleus. Since the reverse of this process, the breaking up of the nucleus, requires an expenditure of energy—in other words, energy must be applied to the nucleus—the nucleus most lose energy in the process of its formation. Therefore, the binding energy of an atomic nucleus is always a negative quantity, by definition. Naturally, the more stable a nucleus is, the more difficult it is to split it into its constituent parts, and the greater is the quantity of work that is necessary for this purpose. Therefore, stability increases with the absolute magnitude of the (negative) binding energy, and thus, strictly mathematically, the smaller the binding energy, the greater is the stability of the nucleus. For this reason, when speaking of a greater or smaller binding energy of a nucleus, we usually refer to its absolute magnitude. In this sense, the greater is the binding energy of a nucleus, the greater is its stability.

Since we are still unfamiliar with the details of nuclear structure, we are unable to calculate the binding energies from the nuclear properties. We must therefore attempt, conversely, to ascertain the magnitudes of binding energies by other methods, in order to use them as a basis for drawing conclusions about the properties of the nucleus.

The simplest example of a nucleus containing more than one particle is the deuteron, the hydrogen nucleus of mass number 2, which consists of 1 proton and 1 neutron. When such a nucleus is formed from these two component parts, the same quantity of energy must be liberated which would be necessary to break it up. Thus we start out from a state in which the proton and the neutron are still in a state of rest, far away from each other, and exert practically no force on each other. Let the quantity of energy present, under these conditions, within the system constituted by these two particles, be called 0. (The choice of zero for the potential energy of a system is arbitrary, and this choice may always be made in the most expedient manner.) As soon as the particles have combined in a deuteron, the energy of the system has decreased by the absolute magnitude of its binding energy. If we manage to measure in some way the energy content of a deuteron, we can ascertain the magnitude of its binding energy from the difference in energy before and after the joining of its constituent parts, and we can use it as a basis for reaching a conclusion concerning the stability of this nucleus.

We can proceed by the same method and add, let us say, another proton. This gives us the binding energy of the helium nucleus $_2He^3$. And we could continue in the same manner, step by step, to ascertain the magnitude of the binding energy of every nucleus.

The physicist customarily measures energy in *ergs* (the erg is a unit of the centimetre-gramme-second system), whereas the engineer measures it in foot-pounds or, e.g. kilowatt-hours. For thermal energies he uses the *calorie* as a unit. Thus, the different branches of physics and technical science use those units, which are most suitable for the field in question because they are designed for the orders of magnitude of energy generally encountered in that particular field, so that the energy units are neither so small nor so big as to be unwieldy or inconvenient to work with. The same principle applies to atomic physics. Atomic physicists often make use of charged particles, electrons, accelerated by high voltage, in order to measure the binding energies of electrons. Therefore, in this case the unit of energy is that quantity of energy which is

THE NORMAL STATES OF ATOMIC NUCLEI 69

gained by an electron—or in general, by any particle which carries one elementary quantum of electric charge—in passing through a potential drop of 1 volt. This energy unit is called the *electron-volt* (*ev*). It is a very convenient unit to use when dealing with extranuclear atomic structure, for it is more or less of the same order of magnitude as the binding energies of this structure. But the binding energy of the nuclear particles is about one million times greater. Therefore, in nuclear physics it is customary to use a million times this unit—1 million ev = 1 *Mev*. One Mev is the energy which a particle carrying one elementary quantum of electricity gains on dropping through a potential difference of 1 million volts. However, it is still a very small quantity when compared with 1 erg, viz:

$$1 \text{ Mev} = 1 \cdot 6 \cdot 10^{-6} \text{ erg.}$$

We have just mentioned the energy which is liberated when a proton and a neutron combine in a deuteron. As a matter of fact, it is possible to measure this energy very well by actually producing this phenomenon. A source of neutrons is required for this purpose. Modern experimental physics has such sources available. Of course, it is true that during the process of their production, neutrons move at very high velocities and must first be slowed down in order to be brought nearly to rest, so as to be capable of combining with hydrogen atoms from this state of rest. We therefore shoot them through some substance which contains hydrogen. The neutrons collide with very many hydrogen atoms, and as a result of these collisions, they gradually lose most of their kinetic energy, except for a very small quantity which corresponds to the temperature of that particular substance. They end up with a so-called thermal velocity. In this state they are caused to combine with protons.

In this process, the binding energy of the deuteron is liberated, and in conformity with the law of the conservation of energy, it must still exist somewhere—in other words, it must go somewhere, in some form. The most logical assumption is that it leaves in the form of an electromagnetic radiation, namely, as gamma radiation of an extremely short wavelength. It is therefore to be expected that the formation of each deuteron, from one proton and one neutron, is associated with the emission of a gamma-ray photon very rich in energy. The relationship of the energy E of this photon to its frequency ν is determined by Planck's formula, $E = h\nu$, where h is Planck's constant. The energy E is thus identical with the amount of the binding energy of the deuteron. This gamma radiation can actually be observed. Since we know how to measure the frequency ν, it is actually possible to measure the binding energy of a deuteron. It amounts to 2·3 Mev. One might surmise that this energy is radiated in the form of several photons, rather than as a single one. However, it can be proved that such a process is actually far more improbable than the emission of one single photon.

There is, however, another, simpler method available for acquiring knowledge concerning the binding energy of atomic nuclei. According to Einstein's theory of relativity, there is a simple relationship between the mass of a body and its energy content. A relationship of this kind was known, in a special form, in pre-relativistic days, too; it follows from the electro-dynamics of moving objects. Hasenöhrl had already pointed out that the radiation enclosed in a cavity had a seemingly inert mass m, proportional to the energy of the system, m being proportional to E/c^2, where c is the velocity of light. But he failed to compute the proportionality factor correctly. The relationship between energy and mass was made clear by the theory of relativity, which—this is the decisive point—raised it to the status of a universal natural law, applicable not only to the theory of radiation, but to every other branch of physics as

THE NORMAL STATES OF ATOMIC NUCLEI

well. In other words, the following relation is true universally:

$$m = \frac{E}{c^2}$$

The meaning of this equation is: Any system having the energy content E has a mass m commensurate with this energy content, and the quantity of this mass is E/c^2. This conclusion has rather strange consequences. For instance, when a clock is being wound, it must become slightly heavier because energy is being stored in its spring. But the quantities of energy in this case are so small as to make the increase in the mass of the clock too minute to be capable of being demonstrated. The mass E/c^2 is much too small in comparison with the mass of the clock itself.

But this relationship between energy and mass can be put to practical use in nuclear physics, where the stored-up energy is of appreciable magnitude compared with nuclear masses. Expressed in the form $E = mc^2$, this relationship enables us to calculate the energy content, E, of a system from its mass, m. The velocity of light is known; it is almost exactly 300,000 km. per second, i.e. 3×10^{10} cm. per second. In atomic nuclei, the orders of magnitude are totally different from those met with in the above example of winding up a clock. It is true that the binding energies of nuclei are very small, but so are their masses too. Therefore, the mass $m = E/c^2$ is no longer negligibly small in comparison with the masses of the nuclei themselves. Consequently, any change taking place in the nuclei as a result of changes in energy content, can be measured with a considerable degree of accuracy. The application of the above formula to atomic nuclei confirms at the same time the important relationship it expresses.

Since energy is liberated when a deuteron is formed from a proton and a neutron, the mass of the deuteron must be less than the sum total of the masses of the proton and the neutron while these two particles still had a separate existence. Thus must apply to every nucleus composed of N neutrons and Z protons. This statement can be expressed in the form of an equation as follows:

$$M_{nucleus} = Nm_{neutron} + Zm_{proton} - \frac{|E|}{C^2}$$

where $|E|$ is the binding energy, taken as a positive magnitude, of the nucleus, which in the case of the deuteron is liberated in the form of a photon.

It is convenient to use the above equation in a slightly modified form, referring it to the neutral atom, i.e. to the nucleus plus the extranuclear atomic structure, as a whole, instead of the nucleus alone. In this case, the hydrogen atom of mass number 1, with its one electron, must be substituted for the proton. As a result, the masses appearing on both sides of equation increase by the mass of the Z electrons possessed by the entire atom, on the one hand, and by the Z protons, on the other. The equation then will read:

$$M_{atom} = Nm_{neutron} + Zm_{H\ atom} - \frac{|E|}{C^2}$$

From this equation we can compute the binding energy of a nucleus if we know exactly the mass of the atom in question, the mass of the neutron and the mass of the hydrogen atom.

The mass of an atom is thus always $|E|/c^2$ less than the sum of the masses of its component parts. This difference in mass is called the *mass defect* of the nucleus. It equals $|E|/c^2$, and therefore:

$$mass\ defect = \frac{|E|}{C^2} = Nm_{neutron} + Zm_{H\ atom} - M_{atom}$$

Atomic masses are usually expressed in terms of the standard physical atomic mass unit (*a.m.u.*) which is approximately equal to the mass of a hydrogen atom, or of a neutron; it is exactly $\frac{1}{16}$ of the mass of the oxygen isotope $_8O^{16}$. The

THE NORMAL STATES OF ATOMIC NUCLEI 73

mass defects are of the order of magnitude of 1/1,000 a.m.u., and therefore it is customary to express them in units of 1/1,000 a.m.u. (a.m.u.⁻³). The energy equivalent of 1 a.m.u.⁻³, i.e. 1 a.m.u.⁻³ × c^2, happens to differ only slightly from the energy unit 1 Mev. To be exact:

1 a.m.u.⁻³ is the equivalent of 0·93 Mev.

Accordingly, the binding energies of nuclei are also of the order of magnitude of 1 Mev.

Thus the physicist has two completely independent processes available for determining the binding energies: He can measure them directly, or can compute them from the mass defects. In the latter case, it is essential, of course, to be able to determine the masses of the atoms to a high degree of accuracy, in order to deduce the binding energy from them, for we are dealing with minute differences, measurable in 1/1,000ths of the masses only. Let it suffice here to say that this is accomplished with the aid of the *mass spectrograph*, an apparatus first developed by Aston. In the mass spectrograph, charged atoms are caused to pass through electric and magnetic fields, where they are deflected. As we have already seen, the extent of their deflections depends on the ratio of their charge e to their mass m, as well as on their velocity. The mass spectrograph is designed so that only particles having a certain velocity are singled out and become the objects of observation. The ratio e/m, can then be determined, and since the charge e of the particles is known, their mass m can be computed—or, to be more exact, their mass, m, can be compared with the mass of the oxygen isotope $_8O^{16}$, which is exactly 16·000 a.m.u. by definition. For the calculation of the binding energies, we need also the exact masses of the hydrogen atom and the neutron. The mass of the hydrogen atom is 1·00813 a.m.u., the mass of the neutron is 1·00895 a.m.u.

Let us now consider, quantitatively, the changes in mass when a deuterium atom is formed. Before the formation of this atom, there was an individual hydrogen atom, $_1H^1$, and an individual neutron, $_0n^1$; after the combination of these two, there is a deuterium atom, $_1D^2$, and a free photon, $h\nu$. This finding can be expressed by the following formula:

$$_1H^1 + {_0n^1} \rightarrow {_1D^2} + h\nu$$

Now the mass of a deuterium atom is 2·0147 a.m.u., whereas the sum of the masses of the hydrogen atom and the neutron amounts to $1·00813 + 1·00895 = 2·0171$ a.m.u. The mass of the deuterium atom is actually smaller than the sum of the masses of its component parts, and this difference is the mass defect, which amounts to 0·0024 a.m.u. (2·4 a.m.u.$^{-3}$). But expressed in terms of its energy equivalent, this mass defect represents a binding energy of about 2·2 Mev, which is exactly the energy released in the form of a photon, as we have already pointed out. Thus we see that there are two totally independent methods for the calculation of the binding energy, and both methods yield the same answer, and therefore provide the best possible proof of the truth of the principle of the equivalence of mass and energy. This is an especially important fact, for here we are not dealing with electric fields, with reference to which this principle was well known even in pre-relativistic times, but with fields of a totally different nature.

The findings just discussed show also why the atomic weights of the elements are not integral multiples of a basic unit, as Prout had conjectured. In the first place, the mass of the proton and the mass of the neutron differ slightly from each other. Furthermore, when they combine, a fraction of the sum total of their masses disappears; this fraction corresponds to their binding energy. So even if the mass of the proton were exactly equal to the mass of the neutron, the atomic weights of the elements would still not be integral multiples of a basic unit. This explains the slight deviations from integral numbers in the atomic weights of the lighter elements. But the very considerable deviations of the atomic weights of the heavier elements must be explained in another way. These deviations are more apparent than real and are due to the fact that the natural elements are, mostly, mixtures of various isotopes. Each isotope consists of atoms with a nucleus of a specific type, the mass number of which is almost exactly an integer. But the average mass number of this mixture of isotopes is not an integer.

Thus we have established the desired criterion for the stability of a nucleus. A nucleus holds together because an amount of work would be required in order to break it up into its component parts. The work required is a measure of the binding energy, which, in turn, is equal to the mass defect of the nucleus, expressed in terms of energy. A nucleus would not be stable if it were possible to break it up without doing any work. We must, however, add that the other conservation laws must be taken into consideration too. According to the law of the conservation of electric charge, an atom cannot undergo any change that would alter the total charge present in the system. Hence, the change of a proton into a neutron, or of a neutron into a proton, cannot take place in the nucleus without compensation. Otherwise, many atoms which are actually considered stable, would have to be unstable. There is, for instance, a boron nucleus having the (not exact) mass number 12, and also a carbon nucleus having the mass number 12. The boron nucleus consists of 7 neutrons and 5 protons, whereas the carbon nucleus has 6 protons and 6 neutrons. Their respective symbols are $_5B^{12}$ and $_6C^{12}$. The mass of the boron nucleus, however, is slightly greater than that of the carbon nucleus. The difference is 0·013 a.m.u., and consequently, the difference in binding energy is about 12 Mev. The carbon nucleus has a greater mass defect than the boron nucleus, so that its component parts are bound together more tightly than those of the boron nucleus. We can assume, therefore, that the boron nucleus is unstable and changes spontaneously into a carbon nucleus. This transmutation would liberate energy to the extent of 12 Mev. But this process can occur only when a neutron of the boron nucleus changes into a proton. However, by virtue of the law of conservation of electric charge, such a process is possible only if the newly formed elementary quantum of positive electricity is compensated for by the simultaneous formation of an elementary quantum of negative electricity and the latter removed from the nucleus. This could take place through the emission of an electron simultaneously with the change of the neutron into a proton. Actually, this boron nucleus is not a stable structure, but a radioactive one. It emits electrons—in other words, negative beta rays—and changes into a carbon nucleus.

Nevertheless, this transformation would not be possible if the law of the conservation of angular momentum did not hold. As we have mentioned before, the angular momentum, or *spin*, of both a proton and a neutron is $\hbar/2$, to be considered positive or negative according to the spatial orientation of the axis of rotation. Therefore, the angular momentum of a nucleus composed of an even number of particles is always an even multiple of $\hbar/2$. Both the carbon nucleus and the boron nucleus consist of 12 particles; consequently, the nuclear spin of each must be an integer multiple of $\hbar$. But pursuant to the law of the conservation of electric charge, an electron must be emitted during the transmutation process, and this electron, too, has an angular momentum (electronic spin) of $\hbar/2$. Therefore, the resulting carbon nucleus would retain an angular momentum (nuclear spin), the value of which is an odd multiple of $\hbar/2$. In this case, therefore, the angular momenta involved would not balance. But in this predicament we must remember that when we studied natural beta radiation, we found that similar difficulties of balancing the energies furnished us with a clue to the existence of the neutrino, ejected simultaneously with the emission of the electron. The neutrino, obviously, accounts for the conservation of the angular momentum. The electrons emitted by the boron nucleus display a continuous series of energy values; this fact leads us to the assumption that a neutrino is emitted simultaneously in this case, too. The facts that the boron nucleus is really unstable and that it eventually changes into a carbon nucleus through the emission of an electron and a neutrino, point to the conclusion that the neutrino also has an angular momentum or spin value of $\hbar/2$, opposite in direction, or mathematical sign, to that of the electron, so that it compensates for the electron.

We have now gained a general view of the conclusions which can be drawn from the three conservation laws with respect to the stability of the nucleus. Our findings may be summed up briefly as follows: The nucleus of an atom will always change spontaneously into another nucleus if this process, firstly liberates energy, and secondly, is compatible with the laws of conservation of charge and angular momentum. It is true, of course, that this spontaneous transformation may occur after the lapse of a considerable period of time only, in other words, that its probability may be very small. But if either one of the conditions just mentioned is not given, the nucleus in question is a stable one.

II. NUCLEAR STRUCTURE

The conservation laws have enabled us to reach far-reaching conclusions concerning the stability of atoms, without resorting to any hypothesis about the conditions within the nucleus or the forces operative in it. Now let us attempt to formulate conclusions concerning the internal structure of the nucleus, on the ground of experiments, independently of specific assumptions about these internal forces. How are the protons and neutrons distributed within the nucleus? Can we compare a nucleus with, say, a drop of liquid, in which the molecules are packed with uniform density? Or should the nucleus be compared rather to a globular star cluster, in the centre of which the stars are extremely close to one another, but more widely separated with greater distances from the centre?

In this connection, the mass defects render a very important service. From them we can calculate the binding energy, and we find that when reckoned for individual particles within the atom, the binding energy is approximately the same in all atoms. The lighter atoms—up to about aluminium—where the absolute magnitude is smaller, are an exception to this rule. In the other atoms, the binding energy of each nuclear particle is always between 6 and 9 Mev. It seems, therefore, that all nuclear particles are bound more or less equally firmly.

We may draw a further conclusion from the size of the nucleus. We can determine, approximately, the diameter of a nucleus from the deflection of alpha particles in the substance in question, by ascertaining what fraction of them shows so strong a deflection that it cannot be accounted for by the effects of the external electric field of the nucleus, but can only be explained by direct collisions with the nucleus. The larger the nucleus, the more frequently will such collisions occur. These experiments have shown, for instance, that the diameter of a uranium nucleus consisting of 238 particles is about four times as large as that of the helium nucleus which consists of 4 particles. Their respective volumes are, therefore, as 1 to 4^3, in other words, the volume of the uranium nucleus is about sixty-four times the volume of the helium nucleus. However, the uranium nucleus has about sixty times as many component particles as the helium nucleus.

These two facts—the approximate equality of the binding energies of the individual particles, and the approximate proportionality of the number of particles to the volume of the nucleus—warrant the conclusion that the protons and neutrons are distributed approximately uniformly throughout the nucleus, because if this were not the case, their binding energies would show considerable differences, both in individual parts of a nucleus and in different nuclear types. Moreover, this fact, in conjunction with the proportionality of the number of particles and the volume, indicates that this density distribution is the same in all nuclei, except, again, for the very lightest atoms. Therefore, we may speak of a homogeneous nuclear substance in all nuclei, which consists of a mixture of protons and neutrons packed with an approximately constant density. There is a slight difference here between different atomic species, with respect to the ratio of the number of neutrons to the number of protons.

Therefore, we obtain a very accurate model of a nucleus by likening it to a drop of liquid. Just as water droplets of different sizes can form from water molecules, so drops of nuclear matter of different sizes—the different atomic nuclei—can form from protons and neutrons. This liquid-drop model has exactly the properties which can be observed in an atomic nucleus. For the molecules are packed with equal density of distribution throughout a drop of liquid, and all the molecules are bound together in it by the same quantity of energy. The knowledge of the existence of a universal, homogeneous nuclear substance facilitates greatly our understanding of nuclear structure.

But this liquid drop has also other, still more subtle characteristics, and now we must investigate whether these have their analogies in the nucleus. Actually, in a liquid drop, not all molecules are bound together equally firmly. The molecules on the surface are linked with the others on one side only, and therefore, on the whole they are less tightly bound than the others. This explains the phenomenon of *surface tension*. Certain considerations concerning energy, which are absolutely analogous to those which we have already applied to atomic nuclei, explain the fact that this surface tension causes these drops to assume a spherical shape. For the surface energy of a drop is proportional to the area of the surface, and it consequently tends to make the surface area as small as possible. We must assume the presence of such surface tension in atomic nuclei, too. As a result of the lesser cohesion of the particles situated on the surface, this surface tension must produce a decrease of the total binding energy, and consequently, of the average energy per particle as well. Just as in a liquid drop, the surface tension is the cause of the spherical shape of atomic nuclei, too.

There is, however, one essential difference between nuclear matter and a liquid. The liquid consists of molecules which are electrically neutral, whereas the nuclear matter contains not only neutrons but electrically charged protons, too. Therefore, our analogy must be one between nuclei and liquid drops containing electrically charged molecules with forces of repulsion operating between them. There is also an electric force of repulsion present in the nuclei of atoms.

III. THE THREE TYPES OF NUCLEAR ENERGY

We may therefore regard the energy content of a nucleus as the sum total of three components. The major part of this energy comes from the nuclear forces which make the cohesion of the nucleus possible. This energy is modified by the surface tension. Finally, a part of this total energy originates from the force of electric repulsion. Let us consider these three components separately and correlate them with the number of protons and neutrons on the basis of a study by v. Weizsäcker.

We shall begin with the nuclear forces. These are the forces which bind the protons and neutrons together, and as we have seen, they are related to the fact that nuclei are capable of emitting both electrons and positrons. This phenomenon is evidently quite symmetrical as between protons and neutrons. A neutron can change into a proton, in which process an electron is emitted, and conversely, a proton can change into a neutron, in which case a positron is the by-product of the process. This fact warrants the conclusion that with respect to the nuclear forces, or to the *nuclear field*, there is no difference between protons and neutrons. Therefore, it must be possible

THE NORMAL STATES OF ATOMIC NUCLEI 81

to represent that part of the binding energy which is the product of the nuclear forces, by some *symmetric* function of the number of neutrons and protons. Now, if we first write this function in a general form, and then, near the point where the number of neutrons is equal to the number of protons, we stop with the second term, we obtain a simple equation for the binding energy per particle (in so far as this energy originates from the nuclear field), as follows:

$$\frac{E_v}{N+Z} = -A + B\frac{(N-Z)^2}{(N+Z)^2}$$

where E_v is that part of the total binding energy which originates in the nuclear field, and which is proportional to the volume. A and B are constants. $E_v/(N+Z)$, the binding energy per particle, is therefore equal to a constant $-A$ when $N = Z$, i.e. when the number of protons is equal to the number of neutrons. But if N and Z are not equal to each other, small divergences occur and in that case the simplest symmetrical function of N and Z is the expression $(N-Z)^2$. But since nuclear matter is homogeneous, the binding energy per particle can depend only on the N/Z ratio, which latter is obtained through division by $(N+Z)^2$. The most general symmetrical function would be, obviously, more complicated. But if it is developed, according to ascending powers of $(N-Z)$, in a Taylor series and broken off after the second term, we obtain exactly the equation given above. This approximation is sufficient, since in the cases of the nuclei observed, N and Z never differ much from each other.

The first term of this equation is negative, as it must be in the case of a binding energy. On the other hand, the second term, which is considerably smaller, is positive, and therefore produces a reduction of the absolute magnitude of the negative binding energy. The latter decreases when N and Z are different from each other. Therefore, from the point of view of energy economy, the most favourable situation seems to be present when a nucleus contains exactly as many protons as neutrons.

But here we must mention a correction depending on the surface tension. The particles situated on the surface are bound together less tightly than those in the interior, and this circumstance, consequently, calls for the addition of a further positive term on the right-hand side of our equation. For since the principal term, $-A$, is negative, this side of the equation is essentially negative, and something positive must be added to it in order to decrease its value. The change in the quantity of binding energy due to surface tension is proportional, in every instance, to the number of the particles situated on the surface, and hence to the surface area. But the latter is proportional to the nuclear volume (or to the total number of particles) raised to the power of 2/3. This portion of the binding energy can therefore be expressed by the equation $E_0 = C(N+Z)^{2/3}$. The portion per particle is obtained by division by $N+Z$, i.e.:

$$\frac{E_0}{N+Z} = C(N+Z)^{-1/3}$$

where C is again a constant.

Finally, another term, a result of the electrical repulsion of the protons, must be added. Here we are in the familiar territory of electrostatics. The charge on a nucleus is Ze, where e is the elementary quantum of electricity. The electric energy of a nucleus is proportional to the square of its charge, i.e. to Ze^2 just as the energy of a condenser is proportional to the square of its charge. Furthermore it is inversely proportional to the radius, r, of the nucleus. There is also a numerical factor to be considered, which for a homogeneously charged sphere is 3/5. If the charge is displaced slightly toward the surface, this factor becomes smaller and will approach $\frac{1}{2}$. Since the radius is known, anyway, not very exactly, we can retain the factor 3/5 despite the fact that there is undoubtedly a certain displacement of the

THE NORMAL STATES OF ATOMIC NUCLEI

charge toward the surface. This energy component may therefore be written:

$$E_c = \tfrac{3}{5} \frac{(Ze)^2}{r}$$

But since the radius r is proportional to the cube root of the volume, i.e. to the cube root of the number of particles, we can write: $r = r_0 (N + Z)^{1/3}$, where r_0 is a constant which would, so to speak, be the radius of a nucleus containing one single particle, but naturally must not be identified with the radius of the proton or the neutron. Thus, this portion of the binding energy per individual particle is:

$$\frac{E_c}{N+Z} = \tfrac{3}{5} \frac{(Ze)^2}{(N+Z)^{4/3} r_0}$$

We must also add this term to our equation with a positive sign, since the electric repulsion reduces the quantity of the total binding energy. The complete expression of the binding energy per particle, finally, is:

$$\frac{E}{N+Z} = -A + B\frac{(N-Z)^2}{(N+Z)^2} + \frac{C}{(N+Z)^{1/3}} + \tfrac{3}{5}\frac{(Ze)^2}{(N+Z)^{4/3} r_0}$$

This equation contains four constants, A, B, C and r_0, of which, to start with, we know only r_0—and even this one rather inexactly—from the measurements of nuclei.

In order to make use of this equation, it is essential to know the four constants exactly. If we knew more about the interior nuclear forces, and about nuclear forces in particular, we could calculate them theoretically. Actually, however, only the converse method is available to us—their empirical determination on the ground of the data already known to us about the binding energies of nuclei, in other words, about mass defects. For instance, a study by Flügge and v. Droste has resulted, by this method, in the following numerical values: $A = 0.01574$ a.m.u.; $B = 0.022$ a.m.u.; $C = 0.0165$ a.m.u.; $\dfrac{3e^2}{5r_0} = 0.000646$ a.m.u.

Thus the binding energies of the nuclei can actually be represented in good agreement with empirical findings. The curve representing our equation, shown in Figure 11, agrees approximately with the above-mentioned values of the constants; this curve shows the energy of the most stable elements for the atomic weights indicated. Since the binding energy is a negative

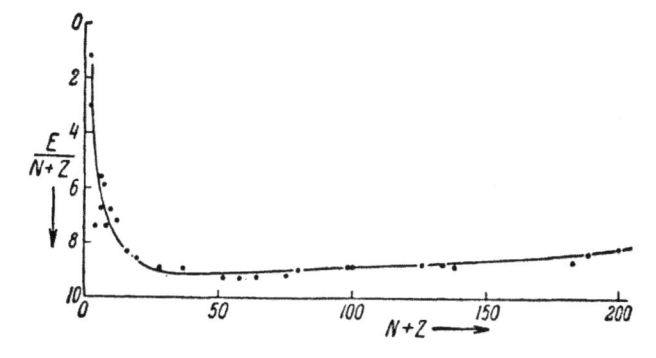

Figure 11.—Binding energies of nuclei as functions of $N + Z$.

quantity, the lower a point is, the greater is the binding energy of the element represented. The dots appearing along the curve are the values of binding energy per particle actually calculated from the mass defects. It will be seen that the agreement is very satisfactory.

Let us now analyse this curve. The position of its lowest point is certainly determined to a dominant extent by the largest term in our equation, $-A$. The rise observed for small atomic weights is due to the surface tension, which naturally plays the most important part for light nuclei. The rise for heavy nuclei is due to the electric repulsion of the protons.

Furthermore, our equation indicates that the lighter nuclei

THE NORMAL STATES OF ATOMIC NUCLEI 85

mostly contain approximately as many protons as neutrons. The last term, which is due to the force of repulsion, increases with the increase in the value of Z, and plays no important part as yet in the cases of light atoms, where Z is small. In such cases, it may therefore be disregarded. The third term is determined solely by the sum total of the number of particles, and not by their ratio, as in the second term, which disappears when $N = Z$, in other words, when the number of protons equals the number of neutrons, and in which case the quantity of binding energy reaches the maximum. This means that from the point of view of energy, this is the most favourable situation. But this is not the case with the heavier atoms, where the last term produces a perceptible decrease of the binding energy. As regards the energy situation, it is more profitable that the second term should increase slightly, in consequence of N/Z slightly exceeding unity, in order that a very considerable diminution in the fourth term may thereby be achieved.

The binding energy is, in fact, a function of both N and Z, and we can represent it as such in a three-dimensional co-ordinate system, recording the number of protons on the horizontal axis running from left to right, the number of neutrons on the horizontal axis running from the front to the rear, and the binding energy on the vertical axis. Our equation will then produce a surface, for a particular energy value belongs to each pair of magnitudes (N, Z). The shape of the energy surface supplies all the information necessary for determining the stability of the nucleus. Figure 12 shows this surface, in the form of a map, with contour lines representing equal magnitudes of binding energy. The energy unit used is 0·001 a.m.u. Since the binding energies are negative, the surface lies below the plane of the drawing, like a valley dropping away from the lower left corner of the drawing toward the upper right corner (in other words, from the south-west to north-east, to keep the map analogy). The stable nuclei lie at the bottom of this valley, like houses along a gently winding street. Those nuclei which contain the same number of particles, always lie on straight lines, running from north-west to south-east at an angle of less than 45 deg. Among these nuclei, the one situated nearest to the bottom is always the most stable one. In so far as it is compatible with the conservation laws, one of the less stable nuclei can always change into a more stable one. The nuclei lying on the left side

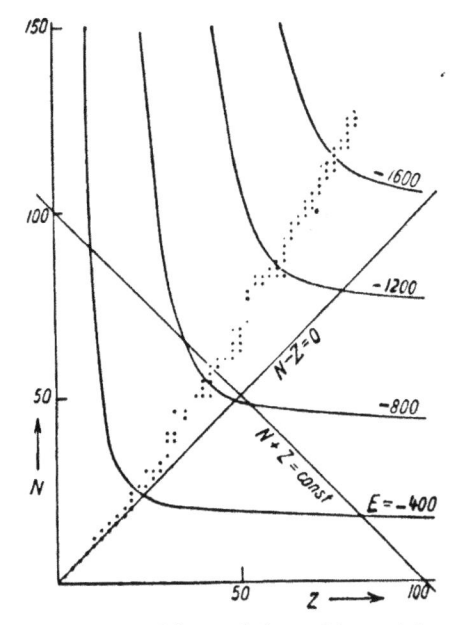

Figure 12.—Positions of the stable nuclei and curves of constant binding energy.

of the valley floor have too many neutrons, so that they must change by the emission of an electron. Those on the right side have too many protons, and would have to change by the emission of a positron. We find, however, that this very often does not happen, and there frequently exist two or three stable nuclei having the same $N + Z$ total. These nuclei are called *nuclear isobars*. This situation can be explained by certain finer details of the valley bottom only, not immediately evident from

THE NORMAL STATES OF ATOMIC NUCLEI 87

our equation based on very general assumptions. Actually, the valley bottom shows small convolutions and other fine details, on which actual measurements of the binding energies will supply information. We may say, nevertheless, that the stable nuclei will lie at the valley bottom or in its most immediate neighbourhood; we can also name the most stable of those nuclei which contain a certain given number of particles.

Figure 13 shows a picture of the situation in a different light. In this case, the combined totals of the numbers of particles,

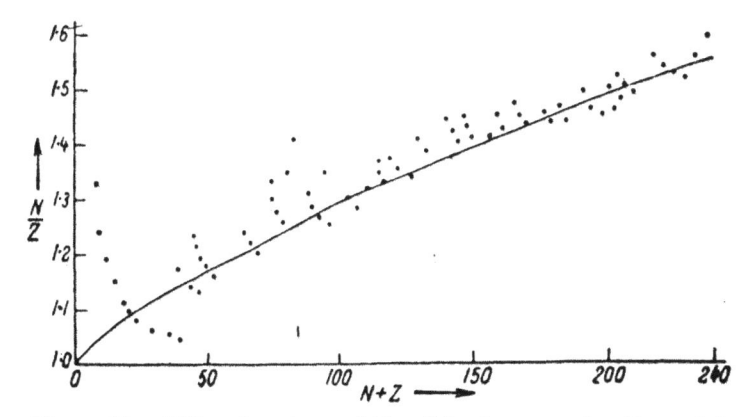

Figure 13.—N/Z as functions of $N + Z$ in the case of stable nuclei.

$N + Z$, appear as abscissae, and the ratios of the particles, N/Z, as ordinates. The continuous straight line corresponds to the dots along the valley floor, and the dots indicate the positions of the individual stable nuclei. We can see how they are distributed along the valley floor and in its immediate vicinity.

In Figures 11, 12 and 13, only stable nuclei are shown. The complete picture is presented by Tables IVa and IVb (at the end of the book) which include all the known unstable nuclei. The abscissae show the number of protons, the ordinates show the difference between neutrons and protons, $N - Z$. Stable nuclei are indicated by black dots, electron emitters by upright triangles (the apex pointing upward), and positron emitters by inverted triangles. Since the electron emitters are relatively rich in neutrons, whereas the positron emitters contain relatively few, the former appear mostly in the region where $N - Z$ is larger—i.e. on the upper periphery of the group of nuclei—while the majority of the positron emitters are found at the lower edge of the group, where $N - Z$ is smaller. It is a striking fact that not infrequently unstable nuclei are also found interspersed among the stable ones. The reasons for this circumstance will be discussed later. Certain nuclei change by capturing an electron from the innermost electron shell (the *K-shell*) of their extranuclear structure. These nuclei are referred to as *K-capturers*, and are indicated in our tables by circles. We find also a number of squares, mostly at the end of the table, indicating nuclei which emit alpha rays.

Up to this point, we have discussed nuclear stability in its relation to transmutation by the emission of electrons or positrons. Let us now briefly discuss stability in its relation to transmutations by alpha ray emission. Studying Figure 11, we can conclude from the initial decrease and subsequent increase in the magnitude of the term $E/(N + Z)$ that nuclear stability first increases steadily, up to about where $Z + N = 40$, but thereafter it gradually decreases, because of the increasing effect of electrical repulsion. Nevertheless, even in these cases, work has to be done in order to remove an individual particle from the nucleus. Let us assume that after applying the necessary work, we remove simultaneously two neutrons and two protons from the nucleus and then combine them in a helium nucleus. In this process a very large quantity of energy, 30 Mev., is liberated. If this energy is greater than that quantity of energy required in order to separate four individual particles, the net result is an actual gain in energy. Therefore, from the point of view of energetics, this process is an advantageous one and ought to be taking place spontaneously, in the form of the formation of an alpha particle in the nucleus and its subsequent emission by the latter. The probability of the occurrence of this process is bound to increase with the increase of the number of particles, for the binding energy per particle decreases as the total number of particles increases. Therefore, alpha emitters are to be looked for among the heavier nuclei, in conformity with actual experience. Actually, as we approach the heaviest nuclei, the magnitude of binding energy per particle drops—approximately at least—to a value between 6 and 7 Mev., in other words to about one-fourth of the binding energy of a helium nucleus.

When the total number of particles is high, the splitting of a nucleus into two parts, not very different in size, is advantageous from the point of view of energetics. A nucleus of mass number 230 might split into two nuclei, one of mass number 100 and another of mass number 130, for the sum total of the binding energies of these two nuclei is greater than the binding energy of the original single nucleus of mass number 230. Instances of such splitting—*fission*—of atoms were actually observed by Hahn and Strassmann in 1938.

As a matter of fact, we might well wonder why it is that not every one of the heavier elements is an alpha emitter or splits into two nuclei of more or less equal size, instead of holding together, as they do according to actual experience, for a remarkably long time at least. As regards fission, the lifetimes of these elements are even longer. We shall inquire into this question in our sixth lecture.

5. THE NUCLEAR FORCES

I. GENERAL PROPERTIES OF THE NUCLEAR FIELD

The cohesion of protons and neutrons within the atomic nucleus is ensured by forces, to which we have referred as *nuclear forces*, although we have not yet discussed their nature. The electric forces of repulsion, which are also operative in the nucleus, exert a purely disruptive effect. What can we learn today from the experiments on the nature of these nuclear forces? Let us first study briefly the form which an answer to this question must take. Assuming that we did not as yet know what electric forces were, how could we begin to explain their nature? We could begin by stating that electric charges mutually repel each other, with a force which decreases inversely as the square of the distance between them. According to the knowledge gained in the early part of the nineteenth century, we could add that a fundamental relationship exists between electric and magnetic forces—e.g.: varying electric forces always generate magnetic forces, and vice versa. Moreover, we might add that the phenomena of light, which used to be regarded as something of a special nature, are among these electromagnetic phenomena, and are simply nothing but electromagnetic waves. The next step would be to note that in certain experiments light appears not as a wave, but as flying particles, in other words, in the shape of photons, and thus we would discover a relationship between the electromagnetic field and these photons. A truly exhaustive description of electromagnetic forces can, however, be given solely with the aid of mathematical equations, expressing how electromagnetic forces change and spread. A complete picture of the 'nature' of these phenomena is supplied only by Maxwell's equations in combination with the equations of the quantum theory.

But as for the nuclear forces we have not got to them yet.

THE NUCLEAR FORCES

Nevertheless, we can already form a picture of them, which is correct qualitatively at least and includes as many details as the picture of the electromagnetic forces—except for the exact mathematical formulae.

The first question to arise here is: How does the force operating between two particles in the atomic nucleus depend on their distance from each other? Is it perhaps also in inverse proportion to the square of this distance? The simplest object available for the study of this problem is the deuteron; we inquire about the force which binds a proton and a neutron in a deuteron. Once this force is known, we have a good prospect of comprehending the cohesion of other nuclei. The force in

Figure 14.—Deflection of a neutron in the neighbourhood of a proton.

question cannot possibly be electrical in character, if for no other reason than that the neutron carries no electric charge. Furthermore, electric forces would be much too weak to account for the considerable energies which result from the mass defects.

We have already pointed out that when a deuteron is formed out of a proton and a neutron, the binding energy is liberated as a photon, with an energy of 2·2 Mev.—in other words, as electromagnetic energy. This means that a process is taking place in which energy is converted from one form to another— the non-electromagnetic energy of the nuclear field into the electromagnetic energy of radiation. Therefore, it follows that in common with all other types of energy, the energy of the nuclear field possesses the capacity of being transformed into other forms of energy.

We can gain some insight into the dependence of nuclear forces on the distance by observing the deflection of flying neutrons when they pass near a proton (Figure 14). Modern physics has access to sources of neutrons. All that is necessary is simply to send the neutrons through a hydrogen-containing substance, for example, a hydrocarbon, such as paraffin, or water, to cause them to be deflected from their straight paths. The magnitude of the deflection of a neutron depends, naturally, on the distance of its path from the proton. This distance is more frequently a greater than a smaller one. Instances of neutrons passing close to a proton are very rare. If the forces decrease relatively slowly with the distance—as, for instance, in the case of electric charges—even though the distance may be considerable, the neutrons will still be deflected a little. It would indeed be observed that a very large number of neutrons are deflected, but the deflection is always very slight. As a matter of fact, large deflections are seldom observed. If, on the other hand, the force diminishes rapidly with the distance, the majority of the neutrons are not deflected at all. In the case of the deflected neutrons—those few which pass sufficiently close to the proton—both small and large deflections may be observed to occur with comparable frequency.

Such experiments have demonstrated that the force between neutron and proton diminishes with the distance more rapidly than is the case with electric forces of attraction and repulsion. The degree of accuracy of the measurements does not as yet permit us to formulate exactly the law of distance. Nevertheless, we can state that the force is becoming already very small at a distance of 5×10^{-13} cm. This means that the force between proton and neutron has an extremely short range, and in this respect it is very different from an electric force.

Instead of studying the force, we may base our considerations on the potential energy which a neutron has within the field of a proton (or vice versa). When the distance is a very large one, we arbitrarily assign to this energy the magnitude 0.

At finite distances, it has a negative magnitude. Due to the limited range, the potential energy is practically 0 for any but a very small distance. Figure 15 shows an approximate curve of the potential energy as a function of the distance r. This energy increases rapidly from high negative magnitudes until it reaches the vicinity of 0, which it then approaches asymptotically. Dealing with small distances, the potential energy curve can be computed indirectly from the mass defect of the deuteron. In addition to its potential energy, the system has also a kinetic energy, since the proton and neutron reciprocally vibrate with reference to each other, with a continuous transformation of kinetic energy into potential energy, and vice versa. The sum total of these energies always equals the binding energy, 2·2 Mev., which is shown in Figure 15 as a horizontal line. The average magnitude of the kinetic energy can be estimated, for instance, in accordance with the uncertainty principle, from the diameter of the deuteron. The accuracy of our knowledge of the exact location of this object of finite diameter goes hand in hand with a proportionate inaccuracy of our knowledge of its velocity; and the squared velocity times one-half the mass gives us the magnitude of the average kinetic energy. Once we know both the kinetic energy and the total energy, we can calculate from them the potential energy. Figure 15 is the result of considerations of this nature.

Short-range forces are known to exist in nature elsewhere, too; the most important example is exhibited by the chemical forces, the so-called *valencies* (unless we are dealing with what is called a *polar* combination), in other words, the forces, which, for instance, bind two hydrogen atoms to one oxygen atom in a water molecule. These, too, are short-range forces which are operative actually only when the atoms are in direct contact, but become infinitesimally small as soon as the distance between the atoms increases.

It is due to this extremely short range that in heavier, macroscopic structures we can perceive neither the chemical nor the

94 NUCLEAR PHYSICS

nuclear forces, whereas electric or magnetic forces are perceptible without any difficulty. The force between two magnetic poles is felt directly by the hand in which a magnet is held, and your hair stands on end as you approach an electrical high-tension apparatus. But chemical forces can never be perceived in this direct fashion, for they are operative over molecular distances only. The same applies to nuclear forces; they cannot

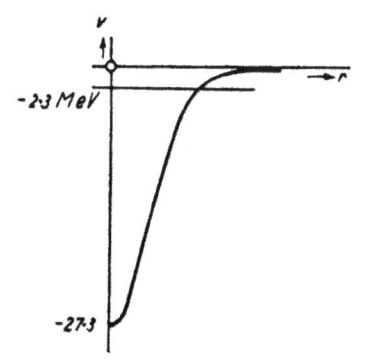

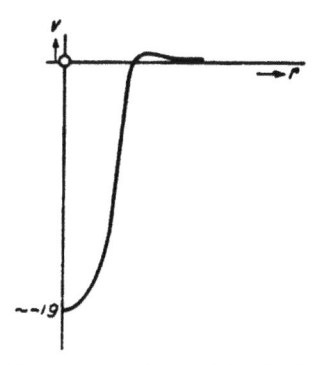

Figure 15.—Potential of the force between neutron and proton.

Figure 16.—Potential of the force between proton and proton.

be perceived anywhere except in the nuclear phenomena themselves.

This has already supplied us with a certain general view of the nature of the force operative between a proton and a neutron. But what about the force between two protons?

One might surmise, to start with, that electric repulsion only is operative between them, since the force between protons and neutrons would be sufficient in itself to explain both nuclear cohesion and the fact that nuclear matter in a stable state consists of an approximately equal number of protons and neutrons. For if some force were operative between protons and neutrons only, the symmetry would already be guaranteed, to begin with. But experience concerning the deflection of protons by protons proves that forces of attraction are acting between particles of the same kind—in other words, not only between protons but between neutrons, too—which forces of attraction are approximately equal to those acting between protons and neutrons. In the case of two protons the situation is more complicated, because the electric force of repulsion is superimposed on the nuclear force of attraction. But when dealing with very short distances, the force of repulsion is much weaker than the nuclear force, so that in this case, practically, only the latter is operative. However, due to its long range, the electric force continues to be perceptible long after the nuclear force has ceased to be operative. If we draw a diagram of the potential energy of a proton at various distances from another proton, it will look more or less like the one reproduced in Figure 16. In fact, up to a distance of the order of 5×10^{-13} cm., the picture is practically identical with the one shown in Figure 15. From that point on, however, the potential energy does not approach 0 asymptotically, but passes through 0, rises to positive magnitudes, and only then does it drop asymptotically toward 0. Between two protons there exists what we call a *potential barrier*, a specimen of which will be discussed repeatedly later on.

According to all this, one might surmise that there exists a state in which two individual protons are bound to each other, namely, when their distances from each other are so small that the nuclear force of attraction overcomes the electric force of repulsion. But this is probably never the case. As pointed out before, two particles bound to each other vibrate constantly in relation to each other, even in their normal or ground state, the state of least energy. This 'zero-point vibration' is probably so powerful that it makes it impossible for any permanent bond to exist between two individual protons. But the attraction between protons is certain to play an important part in the more complex nuclei.

So now we have obtained our first overall view of nuclear forces. The most important of these is the force of attraction between neutron and proton. There is, furthermore, a force of a similar order of magnitude acting between two protons or two neutrons. The operative range of these nuclear forces diminishes rapidly with the distance, and in this respect they resemble the chemical valency forces, which likewise possess a very short range only.

II. THE NUCLEAR FORCES AS EXCHANGE FORCES

Let us now continue to formulate our queries in the same way as we did when dealing with the electric forces. Thus our first question will be: Does there exist any analogy permitting us to link nuclear forces with particles in a way similar to that in which we link electric forces with photons? With this object in view, we must once again study the chart appearing on page 60. The building blocks of the extranuclear structure of the atom are the electrons which are bound to the nucleus by the electric field. The electric field, in turn, is linked with the photons emitted by the atom when certain changes occur in this extranuclear structure. The building blocks of the nucleus are the neutrons and protons, which are held together by the nuclear field, while in this case the electric field is not a binding but a disruptive factor. Here, too, there are particles emitted by the nucleus as a result of changes in state, and in this case, we must distinguish between different kinds of particles. First, there are the gamma rays or photons. Analogously to the photons originating in the extranuclear atomic structure, these photons are linked to the electric field in the nucleus. In addition to these, there are the electrons and positrons, emitted in nuclear transmutations, and the neutrinos which always accompany them. The latter are similar to photons in many respects. The only difference is that neutrinos have a spin or angular momentum of $\hbar/2$, whereas that of a photon is either 0 or $\hbar$.

It seems logical here to assume that the emission of these particles is linked up with the nuclear force field, somewhat in the same way as is the emission of photons with the electric field in the extranuclear atomic structure. But such an analogy would mean that the force between neutron and proton is transmitted because of the electrons, positrons and neutrinos. A similar linking of certain particles to a field must not be misinterpreted as meaning that the field is composed of such particles. The expression 'composed of' always suggests that the field might be conceived, to be, as it were, replaced by such particles. Actually, however, field and particles are, so to speak, merely different aspects of the very same concept, as was discussed earlier, in connection with the extranuclear structure of the atom.

The most correct way of expressing the situation is: There is a nuclear field, and in stationary states this nuclear field takes on the aspect of a short-range field, continually diminishing in intensity away from its centre, while in non-stationary processes it takes on the aspect of a wave radiation. The latter can be observed either as a wave radiation or as particles, according to the method of observation employed. We shall attempt to explain this by comparison with the more familiar electric field, by describing the force exerted by one electron upon another in two languages—first, in the language of waves, and then in the language of particles.

We can say, first, that an electron produces an electric field around itself, and that this electric field spreads in conformity with Maxwell's equations. It may act on another electron and create a force on the latter. The corresponding description in terms of the other aspect is: One electron produces a particle, a photon, and this photon is subsequently absorbed by another electron. Thus in the first phrasing we speak of a 'production of a field,' and in the other of a 'production of a particle'; in the first statement, we refer to an ' action of a field', in the second one to an 'absorption of a photon by a particle'. This state of affairs can be expressed schematically as follows:

Wave Aspect: Electron creates field; field acts on another electron.

Particle Aspect: Electron emits photon; photon is absorbed by another electron.

Both statements describe the same event. The first version is familiar to everybody who has ever had any dealings with electric fields. The second one is unfamiliar to most people, because in technical science as well as in macroscopic physics it is always unnecessary to conceive of an electric field as linked to photons. Under atomic conditions, however, this very frequently proves to be a useful expedient. With reference to atomic radiation, it is often more convenient to speak of photons than of spherical waves.

Now let us apply exactly the same type of phraseology to the forces operative between protons and neutrons. First, we can say: The neutron produces a nuclear field, and this field acts on the proton. This is the description in terms of the wave aspect. In the terminology of the particle aspect, our description will be: The neutron produces particles, and these particles are absorbed by the proton. Let us again express this schematically, as follows:

Wave Aspect: Neutron creates field; field acts on proton.

Particle Aspect: Neutron emits electron plus neutrino; electron and neutrino are absorbed by proton.

Interpreting in this manner the force operative between neutron and proton, we see that an exchange of charge is linked with the action of the force. Namely, if in order to exert this force, the neutron must emit an electron and a neutrino, its charge is altered; it changes into a proton. And conversely, a proton changes into a neutron, due to the absorption of an electron and a neutrino. An exactly analogous conversion may occur also when a proton emits a positron and a neutrino, which are then absorbed by the neutron.

Thus, the nuclear forces are associated with an exchange of charged particles, and for this reason, forces of this kind are called *exchange forces*. They are of a very peculiar character, and it is their characteristic feature that their action is linked

THE NUCLEAR FORCES

with an exchange of roles between the two partners. In this respect, therefore, they are totally different from the electric forces. But a close relationship with chemical forces is again evident. Quantum theory has already shown that chemical forces may also be regarded in general as exchange forces. For a similar exchange of charges occurs in the case of chemical forces, too. The simplest example of this is the hydrogen molecule ion, which consists of a hydrogen atom and a hydrogen nucleus (Figure 17). It is therefore actually a structure composed of two protons with an electron circling around them. This ion is a truly stable structure, and the force which holds it together owes its existence to the circumstance that the one electron revolves at times around one proton, at times around the other one. This means that in this case, too, we find that the force is linked with an exchange of charge—the shift of the electron from one proton to the other.

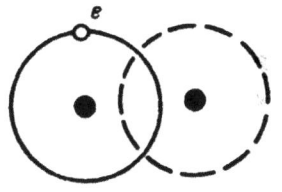

Figure 17.—Ion of hydrogen molecule.

The concept of 'exchange force' can be comprehended most easily on the ground of the following experiment, carried out with the big Lawrence cyclotron in California in 1948 (cf. page 155). Neutrons of great energy are hurled against protons in a cloud chamber. The paths of the protons after collision become visible in the cloud chamber. In the case of an ordinary force, one would expect the great majority of the neutrons to be deflected to a very slight extent only (since they would not hit the proton exactly in the centre), while the protons would be hurled aside, with a relatively low velocity, at an angle of 90 deg. to the path of the neutrons. But in the case of an exchange force, the neutrons and protons must exchange roles after the collision; in the majority of collisions, the protons must continue along the paths of the oncoming neutrons (since the neutrons have actually been changed into protons)

100 NUCLEAR PHYSICS

while the neutron is hurled aside at an angle of approximately 90 deg. This is exactly what we actually see in the cloud chamber. Figure 17A shows the tracks of the protons, most of which fly on along an almost straight path in the direction in which the oncoming neutrons were travelling. A magnetic field

Figure 17A.—Scattering of neutrons in hydrogen.

causes their paths to become more or less strongly curved, according to their energy. (The straight lines slanted at less than 45 deg. are produced by a thin wire mesh in the cloud chamber.)

The situation is, however, not quite so simple as we have made it appear. If the analogy of nuclear and electric forces were as we have been supposing, we would be in a position to determine the frequency of the occurrence of beta decay in a manner similar to that employed to determine the frequency of the occurrence of the emission of a photon in the extranuclear structure of the atom. When an extranuclear atomic structure is in an excited state, there exists in it at any given moment a certain probability of the emission of a light ray. By 'certain probability' we mean the following: In the wave aspect, the continuous movement of electrons causes a wave radiation to issue forth. In the particle aspect, there exists at any given moment a certain probability of the emission of a photon. These two views of the situation are linked to each other by the fact that the probability of radiation is given by the intensity of the emitted wave. The stronger is the wave, the greater is the probability of radiation, and the shorter-lived is the excited state. The duration of the excited state depends, therefore, on the amplitude of the vibration of the electrons.

The lifetime of a beta-unstable nucleus thus depends on the intensity of the wave radiation issuing forth from it. But if we carry out this computation on the ground of the considerations outlined above, we arrive at lifetimes much shorter than those actually observed. There still exists a discrepancy at this point, and this realization led the Japanese scientist Yukawa to a somewhat modified theory.

Yukawa assumes that between the nuclear field and the electrons, positrons and neutrinos there is still another species of particle, which may be called *Yukawa particle* for the time being. These Yukawa particles are assumed to have a mass several hundred times that of an electron, and to be capable of disintegrating into electrons, positrons and neutrinos, directly or eventually through other decay processes. So, according to Yukawa's theory in nuclear transmutations, such a Yukawa particle should actually be emitted. However, this does **not** happen, because the Yukawa particle has such a great rest mass that the energy, mc^2, necessary for its formation is not available. But the Yukawa particle can break up (directly, or indirectly through other processes) into electrons and neutrinos, and this happens, in a certain sense, in the moment of its formation, so that on the whole, it is sufficient to provide the energy required for the formation of the light particles, the electron and neutrino. This theory thus regards the process of nuclear transmutation as occurring in several steps. First, the Yukawa particle is formed from the nuclear field—or more correctly, the nuclear field itself is identical with the Yukawa particle, which for lack of sufficient energy for its formation cannot manifest itself as a real particle. Instead, no sooner is it formed than it breaks up into electrons and neutrinos, which then actually leave the nucleus.

If we accept this theory as a working hypothesis—and there is much in it to make it plausible—there arises the question whether the Yukawa particles are perhaps identical with a certain species of particle already observed in cosmic radiation. Actually, the most recent experiments make it extremely probable that the role of the Yukawa particles is played—in part, at any rate—by the heavy mesons (or 'π particles') observed by Powell; for in cases of nuclear fission of very high energy, these π particles have been observed to be hurled forth from the nuclei. The π particles (already mentioned on page 55) are about 275 times as heavy as an electron. According to Powell's observations, they first break up into a light meson and a neutral particle (the latter is probably simply a neutrino). This light meson (its mass is about 213 times the mass of an electron) then breaks up further into an electron and probably two neutral particles. It is recognized here that the emission of electrons and neutrinos can occur by very roundabout ways only, possibly due somehow to the fact that the probability of the occurrence of a beta decay is extraordinarily small compared with the probability of other nuclear changes.

These considerations show also that the problem of the relationship between nuclear forces and the elementary particles regarded as being linked with them is a very complex

THE NUCLEAR FORCES 103

one, which cannot be solved for many more years to come. At this moment, the only thing we know for certain is that the nuclear forces are, to a considerable extent at any rate, exchange forces, and that there exist unstable elementary particles, the mass of which is between the mass of an electron and the mass of a proton, and which are associated somehow with these nuclear forces. Any further clarification will become possible only when the very high energy nuclear disintegration processes have been investigated much more thoroughly than they have been up to now.

III. THE SATURATION OF NUCLEAR FORCES

The above-mentioned analogy with valency forces suggests a conclusion which, even though we cannot prove it here, we can at least make plausible. An essential difference—among others—between valency and electric forces is that the former are capable of saturation. The chemist provides the symbol of every atom with a certain number of valency bonds, the number of which corresponds to the valency of the element. In every structural formula of a saturated chemical compound, such valency bonds are represented by lines issuing forth from the atomic symbol and terminating in another atomic symbol, from which, in turn, as many such lines issue forth as the numerical value of its valency number. For instance, carbon dioxide, O=C=O, the compound of the tetravalent carbon atom and two divalent oxygen atoms. The characteristic feature in this case is that an atom, all the valency bond lines of which are taken up, is saturated, in other words, it has used up its valency. Thus, for instance, in the case of the water molecule, H—O—H, the two valencies of the oxygen atom are saturated by hydrogen atoms, and no further hydrogen atoms can be bound. A more or less normal molecule of the OH_3 type is non-existent.

Now the nuclear forces or at least a large part of them have a completely similar property of saturation. A neutron can bind to itself not more than two protons, and a proton not more than two neutrons. If we wished to express this fact in the form of valency lines, the proton would have to be written with two such lines, which could connect with neutrons only, and similarly, a neutron would have two valency lines, which could connect with protons only. This is, of course, not quite correct, since we are disregarding the forces which act between any two protons or any two neutrons. Nevertheless, this statement furnishes a preliminary understanding of the actual conditions. This property of saturation of nuclear forces explains why—as already discussed—the binding energy of the individual building blocks of nuclei is independent of the size of the nucleus. If a particle becomes lodged in a nucleus, owing to the extremely short range of the nuclear forces it enters into interaction with its immediate neighbours only—quite unlike the case of electric forces—and secondly, owing to the saturation of the forces, this particle is capable of linking itself to two of these immediate neighbours only. This is a further justification of the analogy between nuclear matter and a liquid. For the same applies basically to the atoms in a liquid, owing to the very similar properties of the forces responsible for their cohesion.

IV. THE STABILITY OF NUCLEI

All that we have set forth above points to the important conclusion that the number 2—and in general, the even numbers—must have a preferred position in atomic nuclei. Consequently, we may expect those nuclei in which both the number of protons and the number of neutrons are even numbers to be particularly stable. For since every proton is able to bind two neutrons, and conversely, every neutron is able to bind two protons, in this case—and only in this case—all valencies can be utilized. This situation is, naturally, more advantageous from the point of view of energetics than one where the valency of an odd proton or neutron is not utilized.

But there is still another reason for the preferred position of the number 2: namely, Pauli's exclusion principle, which we discussed when dealing with the extranuclear structure of the atom. Stated in general terms, this principle asserts that in a stationary state (which with reference to the extranuclear atomic structure means a certain definite electronic orbit, or expressed in terms of the wave aspect, a certain constant vibration with a certain given direction of the electronic spin) there is room for one single particle only at any one time. Taking into consideration the fact that the spin of an electron may be either positive or negative—clockwise or counterclockwise—this can be expressed in the following form: No more than two electrons (with opposite spins) can occupy the same stationary orbit of the extranuclear structure of the atom. The same applies to the protons and neutrons in the nuclei, which likewise possess a spin momentum. It follows, therefore, that not more than two neutrons or two protons can occupy the same stationary orbit in a nucleus. From the point of view of energetics, it is of course more advantageous to have the possibility thus allowed fully utilized. This fact again results in a preferred position for nuclei containing an even number of neutrons and an even number of protons.

This preferred status of the number 2 is especially obvious in the common helium nucleus $_2He^4$, which consists of two neutrons and two protons. This is an especially stable structure, and so is its extranuclear electron structure, which consists of two electrons. This is demonstrated by the fact that helium is an inert gas and does not enter into any chemical combination whatsoever. Actually, the binding energy of the helium nucleus is extraordinarily high: roughly 30 Mev. On the other hand, the binding energy of the deuteron, composed of one neutron and one proton, is only 2·2 Mev., as already mentioned. In the deuteron, only one valency of the proton and one of the neutron are utilized, whereas in the helium nucleus all valencies are saturated.

It is therefore to be expected in general that the preferred nature of even numbers will manifest itself in a particularly high stability of the nuclei which contain both an even number of protons and an even number of neutrons, which we shall call 'doubly even nuclei'. Those nuclei in which either N or Z is an even number while the other one is odd, are less stable; those in which both N and Z are odd, are still less stable. There is a general empirical proof of this statement: It might be expected that the more stable a certain nucleus is, the more frequently it would occur in nature, since at the time of the original formation of nuclei from their building blocks, the most stable nuclei would have a preferred status both as to frequency of formation and the chance of remaining intact. Many years ago, Harkins attempted to determine the empirical relationships between odd or even numbers on the one hand and the natural abundance of elements on the other. He found that the by far most abundant atomic species were, in fact, those which today we know to have 'doubly even' nuclei. Nuclei in which either N or Z is an odd number—'odd nuclei'—are much rarer, and rarest of all are the nuclei in which both N and Z are odd numbers, which we shall call 'doubly odd' nuclei.

'Doubly even' oxygen, $_8O^{16}$, is one of the most common elements. The 'odd' lithium, $_3Li^7$, is much rarer. Finally, we must note the fact that only a very few stable 'doubly odd' nuclei exist at all. The simplest one of these is the deuteron, the nucleus of deuterium, $_1D^2$. The only others are: the lithium nucleus $_3Li^6$, the boron nucleus $_5B^{10}$, and the nitrogen nucleus $_7N^{14}$. All the others of this kind are radioactive and change ('decay') by the emission of electrons or positrons.

On the ground of the foregoing considerations, let us now analyse the question of nuclear stability in greater detail. We have already described the energy surface (Figure 12). This is a very steeply slanting surface with a channel or groove, at the bottom of which the stable nuclei lie. We shall now make a cut through the surface, from the upper left to the lower right, at an

THE NUCLEAR FORCES

angle of 45 deg. to the axes, diagonally to the channel, so giving us a cross-section of the surface (Figure 18). As a result of the nature of this cut, this cross-section will contain only nuclei with the same $N + Z$ total, i.e. with the same mass number. We arrange this cross-section first of all so that $N + Z$ is an odd number. The result is a curve, the lowest point of which is on the valley floor. Only those nuclei lie on this curve which are capable of changing into each other by the emission of an

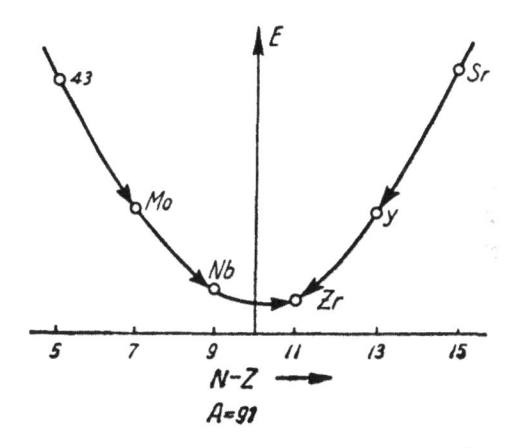

Figure 18.—Binding energy of uneven nuclei.

electron or positron; only the lowest nucleus, the nucleus possessing the greatest binding energy, is expected to be stable. In this diagram, the neutron numbers (their absolute magnitudes are of no importance here) are recorded on the abscissa in ascending order, while the binding energy is shown by the ordinate. Since $N + Z$ is constant, an increase in the magnitude of $(N - Z)$ means an increasing N and a diminishing Z. Those nuclei which lie to the right of the lowest point, have a greater number of neutrons and a smaller number of protons than the stable nucleus, and will gradually change into the stable nucleus by the emission of one or more electrons. On the other hand, the nuclei situated to the left, which contain more protons and fewer neutrons than the stable nucleus, will effect their change by the emission of positrons or by the capture of an electron from the extranuclear atomic structure (*K*-capture), as shown by the arrows in Figure 18. All this is in full agreement with experimental evidence. Figures 18 and 19 show the conditions for the specific mass numbers 91 and 92.

The situation is totally different in the case of 'even' atoms, i.e. where $N + Z$ is an even number. Here one of the finer details of the energy surface (not reflected in the equation formulated in

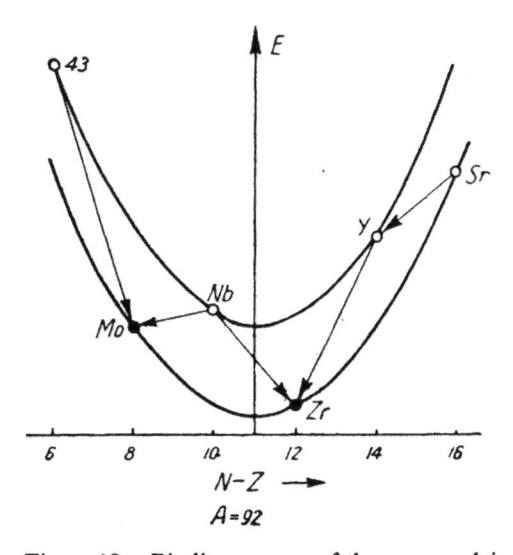

Figure 19.—Binding energy of the even nuclei.

the fourth chapter of this book) shows up, namely: the difference already discussed in the stability, and consequently, in the binding energies, of the 'doubly even' and 'doubly odd' nuclei. As already pointed out, the former possess a considerably greater degree of stability than the latter. Therefore, in order to represent both nuclear types, we must draw two different curves; the first one, representing the 'doubly even'—and therefore the more stable—nuclei lies below the other one which represents the 'doubly odd' nuclei. The change of one nucleus into another can take place by discrete steps only, by the emission of one electron or positron (plus the necessary neutrino) at one time and never by a simultaneous emission of two electrons or two positrons. While it is still possible, for instance, for the nucleus in which $N - Z = 6$ (e.g. $N = 49$ and $Z = 43$— a 'doubly odd' nucleus) to change into the 'doubly even' Mo nucleus in which $N - Z = 8$ ($N = 50$ and $Z = 42$), by the emission of a positron, as shown by the arrow in Figure 19, for energy is liberated in this process. But in order to change further into the most stable Zr nucleus, in which $N - Z = 12$ ($N = 52$ and $Z = 40$), the Mo nucleus with an $N - Z$ value of 8 would first have to change, by the emission of another positron, into the Nb nucleus with an $N - Z$ value of 10—in other words, once again into a 'doubly odd' nucleus. This is, however, impossible from the point of view of energetics, for it would require an expenditure of energy. It is much easier for the Nb nucleus with an $N - Z$ value of 10 to change, conversely, into the Mo nucleus where $N - Z = 8$, by the emission of an electron, or into the most stable Zr nucleus where $N - Z = 12$, by the emission of a positron. Thus, a study of Figure 19 will show that in addition to the most stable nucleus, Zr, situated on the valley floor, other 'doubly even' nuclei with the same mass number, situated a little higher, may also be stable, whereas all the 'doubly odd' nuclei, situated on the upper curve, are unstable. In Figure 19, too, the possibilities of change are indicated by arrows. The arrows pointing to the lower right indicate positron emission (in certain cases, a K-radiation), while those pointing to the left indicate electron emission. Thus, the unstable nuclei situated on the upper curve may change by either process, if there are appropriate stable nuclei on the lower curve. An example is the potassium nucleus $_{19}K^{40}$, which can change by electron emission into the calcium nucleus $_{20}Ca^{40}$, or by positron emission into the argon nucleus $_{18}A^{40}$.

On the ground of the evidence of these curves, we formulate

therefore the following rules, to be verified by actual experiments:

1. When the mass number is an odd figure—in other words, in the case of 'odd' nuclei—there is only one stable nucleus for any given mass number. All the others are unstable and emit either electrons or positrons (or decay by K-capture).

2. When the mass number as well as the number of neutrons and the number of protons are all even figures—in other words, in the case of 'doubly even' nuclei—usually there are several but not many (say, two or three) stable nuclei with the same mass number.

3. Stable nuclei which have an even mass number but an odd number of protons and an odd number of neutrons—the 'doubly odd' nuclei—are likely to be non-existent as a rule.

Exceptions to the third rule mentioned, however, occur among the very lightest nuclei. These can be accounted for by a particularly sharp flexure of the two curves, the result of which is that the most stable nucleus, situated approximately at the lowest point of the upper curve, lies below the nearest nuclei of the lower curve.

Otherwise, all the above conclusions are borne out almost completely by actual experience, as will be evident by reference to Table IV (at the end of the book), in which stable nuclei are indicated by black dots, the unstable ones by upright triangles (electron emitters) or inverted triangles (positron emitters). A careful study of this table will show that, in fact, except for a very few instances, there is never more than one stable nucleus for any given mass number. This statement is known as *Mattauch's rule*. In this table, too, atoms having the same mass number are situated on a straight line, ascending to the left at an angle of 45 deg. to the abscissa. Thus, for instance, the palladium nucleus $_{46}Pd^{111}$, the silver nucleus $_{47}Ag^{111}$, the cadmium nucleus $_{48}Cd^{111}$ and the indium nucleus $_{49}In^{111}$ are all situated on such a straight line. Of these, only the $_{48}Cd^{111}$ is a stable one; the $_{46}Pd^{111}$ and $_{47}Ag^{111}$ nuclei are electron emitters, while the

THE NUCLEAR FORCES

$_{49}$In^{111} nucleus is a positron emitter. Thus this table fully confirms the fact that on the same straight line with a stable nucleus of an odd mass number all other nuclei of the same mass number are unstable. There is an exception to this rule, in the case of the mass number 113. In addition to the stable indium nucleus $_{49}$In^{113} there is also a stable cadmium nucleus, $_{48}$Cd^{113}. This exception can probably be accounted for by the fact that both these nuclei happen to be situated at an approximately equal height on both sides of the energy surface, and the difference in energy between them is too small to permit the formation of an electron and a neutrino, or to admit of any measurable probability of the capture of an electron, so that a change of one of these nuclei into the other cannot occur. There may be also other exceptions, but their existence has not so far been confirmed by experience.

When the mass number is even, the occurrence of several stable nuclei having the same mass number is the general rule. Such nuclei are called *nuclear isobars*. As a result of this privileged character of even numbers, for every nucleus containing an even number of protons (having an even atomic number) there exist more or fewer stable isotopes—i.e. stable nuclei having the same atomic number and the same chemical properties, all of which are varieties of the same element. On the other hand, elements with an odd atomic number have far fewer stable isotopes. Thus, the element titanium, with its even number of protons (22), has as many as five stable isotopes, while its neighbour, vanadium, with its 23 protons, has only one single stable isotope. The next element, chromium, has again four isotopes, whereas the immediately following one, manganese, has just one. Cadmium, with its 48 protons, has as many as eight stable isotopes, whereas silver ($Z = 47$), which immediately precedes it, has only two. And so on throughout the entire periodic system.

The assumptions already discussed concerning the nature of nuclear forces, in particular their short range and their capacity for saturation, are thus definitely confirmed by actual experience in the form of conclusions based on them regarding the relative abundance and stability of the various nuclear species.

This disclosure of the existence of isotopes of the various elements furnishes an explanation of the fact that Prout's hypothesis, based on the assumption that all the known atomic weights are integers—which was in some degree a movement in the right direction,—remained consigned to the limbo of oblivion for nearly a whole century. Subsequent measurements of the atomic weights of the heavier elements proved most of them to be very far from integers, or even approximately integers. But there is a simple explanation of this observation: the existence of isotopes. Every element made chemically pure is a mixture of isotopes (unless it has no more than isotope). The mass numbers of the individual isotopes are, in fact, always approximately integers. But chemical processes only give the average mass of the atoms in a mixture of isotopes depending on the proportions in which the isotopes are present, and which therefore may assume all possible non-integral values.

6. THE NUCLEAR REACTIONS

I. ALPHA RADIATION

Much has been said in the preceding lectures about the changes of atomic nuclei. In such changes, one chemical element turns into another, and in this respect modern nuclear physics has realized, to a certain extent, the hopes which inspired the alchemists of past ages. Let us now consider these nuclear transmutations more closely. The following two questions arise here: What elements can be changed into each other? Under what conditions is such a transmutation at all possible? In order to answer these questions, we will classify the transmutation processes into two groups: Firstly, the reactions which occur spontaneously, and secondly, those produced by external agencies.

The spontaneous change of an element is called *radioactivity*, for the process is accompanied by an emission of radiation. The radioactive elements can be further subdivided into two classes: Firstly, those which emit alpha radiation, and secondly, those which emit beta radiation (either electrons or positrons). Gamma radiation may occur simultaneously, too. In addition to these, there are a few other processes, which will be discussed later.

We shall begin with those transmutation processes in which alpha rays are emitted. Alpha radiation consists of helium nuclei, each composed of two neutrons and two protons. We have already discussed when an emission of alpha radiation may be expected to occur. Approximately, from the element zinc upwards, the binding energy per particle decreases with the increase of the number of particles, due to the increasing intensity of the electrostatic forces of repulsion, and consequently, from the point of view of energetics, the emission of an alpha particle by the heavier elements may be advantageous under certain circumstances. When this occurs, the energy and range of the alpha particle are determined by the difference in mass defect between the original nucleus and the nucleus which is the product of the process. Thus, all alpha particles which are the product of a specific decay process have the same range (Figure 3). Actually, most of these alpha emitters are at the end of the periodic table of elements. Radium and uranium are the best-known ones.

One might be inclined to assume that a nucleus in which the principles of energetics admit the possibility of the occurrence of alpha decay, would disintegrate immediately or at least within a very short period of time. However, that this is not the case is proved by the fact that large quantities of uranium still exist in the world, and that these actually decay only very slowly. In fact, their atoms have existed in an unchanged form for several thousand millions of years. The same applies to thorium and actinium. As a matter of fact, if these three long-lived radioactive elements did not exist, radioactivity would have disappeared from the world a long time ago, except for a few of the lighter elements. For the majority of the other natural radioactive substances originate from these, and are much shorter-lived. The products of the decay of a substance are designated collectively as a *radioactive series*. There are three natural radioactive series: the *uranium series*, the *thorium series*, and the *actinium series*.

It was stated in the third lecture that the *half-life* is used as the measure of the life span of a radioactive substance; the half-life is the period of time during which one-half of the number of atoms present at the start of the period disintegrates. The half-life periods of the various alpha emitters show extraordinary differences in order of magnitude. The half-life of uranium, for instance, is 4,600 million years, whereas the half-life of one of its daughter elements, radium C', is just about one one-millionth of a second. Between these two extremes there are all conceivable intermediate values; for instance, the half-life of radium is 1,580 years. This raises the question as to the cause of these vast differences.

In this connection, it is a very important fact that there exists a simple relationship between the energy of the alpha particles and the half-life of the substance in question. This relationship was discovered, relatively early, by Geiger and Nuttall. These two scientists found that, on the whole, there exists a linear relationship between the logarithm of the decay probability (the reciprocal of the average life of the atom) and the energy of the alpha particles. The latter, as indicated by their equal range (Figure 3), is a magnitude characteristic of every specific radioactive substance. The greater is the quantity of energy available for the decay process, and the greater is the energy of the particle, the more quickly will the decay occur, on the average. If λ designates the decay probability, discussed in the second chapter of this book, and E the energy of the alpha particle, we can write the Geiger-Nuttall relationship in the following form:

$$\log \lambda = A + BE$$

where A and B are constants, to be determined experimentally. The value of the decay coefficient can be obtained from the already mentioned equation $N = N_0 e^{-\lambda t}$, which indicates the number of the atoms not yet decayed during the time t. The energy E can be computed, from the range, by means of a law also discovered by Geiger and Nuttall.

Figure 20 shows the relationship between $\log \lambda$ and E for all alpha emitters, in the form of a diagram based on measurements. We see here three adjacent curves, one for each radioactive series—the uranium series, the thorium series and the actinium series. Although these lines are not perfectly straight, as the Geiger-Nuttall relationship would require, they are at least not too strongly curved. The very fact that we get three different, approximately parallel curves, shows that although the constant B is the same for all the three radioactive series, the

116 NUCLEAR PHYSICS

values of the constant A differ slightly. At the very bottom we find uranium, with the smallest decay probability, and at the very top is radium C' with the highest such probability.

The point now is to explain this regularity, and above all, the fact—very astonishing at first glance—that such a small change in the energy, E, produces such an enormous change in

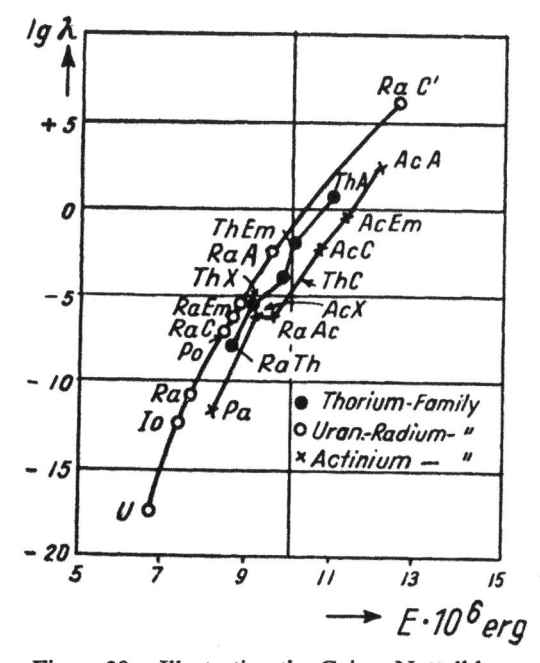

Figure 20.—Illustrating the Geiger-Nuttall law.

the decay probability, λ. In this entire range, the energy varies only between 6×10^{-6} and 13×10^{-6} ergs—a ratio of 1 to 2. On the other hand, the decay probability varies between the orders of magnitude of 10^{-18} and 10^6 per second—a ratio of 1 to 10^{24}.

This fact was explained by Gamow, Condon and Gurney, in 1928. In order to understand this theory, let us first discuss an imaginary experiment. Let us imagine that we have captured an

THE NUCLEAR REACTIONS

alpha particle which has just emerged from the nucleus and have restored it to the place from which it came, and investigate what forces are acting upon this alpha particle, and what work must be applied. The conditions which we encounter are similar to those which prevail in the proton-neutron relationship (Figure 15). As long as the alpha particle is at a considerable distance from the nucleus, it is subject solely and only to the action of the field, by the positive charge of which it is repelled.

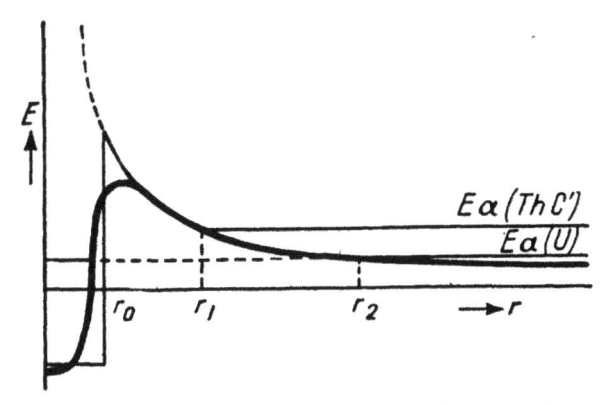

Figure 21.—Potential between heavy nucleus and alpha particle.

Therefore, work has to be applied to the particle in order to bring it closer to the nucleus. This means that as our particle approaches the nucleus, its potential energy increases at first. But when it has come sufficiently close to the nucleus, the short-range nuclear forces of attraction become operative, and they finally overcome the electric repulsion. After a certain potential barrier has been overcome, the force of repulsion changes into a force of attraction, and from there on, the potential energy drops steeply in the direction of the interior of the nucleus. This change in potential energy is shown by the curve in Figure 21.

When the alpha particle is hurled out of the nucleus, it also passes through this potential range, in this case proceeding from inside outwards. Since it reaches a considerable distance from the nucleus with a considerable quantity of kinetic energy, its total energy at that point is a positive magnitude, since its potential energy vanishes. This fact is indicated in Figure 21 by two straight horizontal lines, one of which corresponds to the slow alpha particles of long-lived uranium, the other to the faster alpha particles of the vastly shorter-lived thorium C'. Since the alpha particle carries this energy along out of the nucleus, it must have been in possession of it while still in there. Therefore, we extend the straight line of the uranium right into the interior of the nucleus. The kinetic energy in the interior of the nucleus was obviously still greater; its value at any given point is indicated by the separation of the horizontal straight line from the potential energy curve. The illustration suggests that the particle vibrates to and fro in the interior and bounces backwards and forwards, one might say, from one side of the ' potential container ' to the other. The energy in the interior, shown by the straight line, is always the sum total of its kinetic and potential energies.

At first glance it seems to be impossible to see clearly how this particle can emerge from the nucleus at all. For according to the concepts of ordinary mechanics, it would not be able to move any further outward than the sides of the 'potential container', for at the point where the straight line crosses the potential curve, the kinetic energy vanishes—the particle comes to rest. The energy at its disposal does not enable it to traverse the potential barrier which separates the interior from the outside space. Thus according to classical mechanics, a decay reaction could not take place at all. The potential barrier would guarantee the stability of the nucleus. It could, of course, be assumed that the nucleus contains other particles as well which are likewise in motion, and that these particles might transfer energy to the alpha particle to help the latter cross the potential barrier. But this can occur only when the nucleus is in an excited state—when it has a surplus of energy. In the normal state of the atom, no such free energy is available. For the energy which the particles still possess is—according to the uncertainty principle—zero-point energy, and this energy cannot be utilized, nor can it be transferred to other particles.

This is where wave mechanics comes to the rescue. The movement of alpha particles is governed by the laws of wave mechanics and quantum mechanics, not by the laws of ordinary mechanics. On the ground of the repeatedly mentioned wave-particle duality, we can think, instead of a particle vibrating

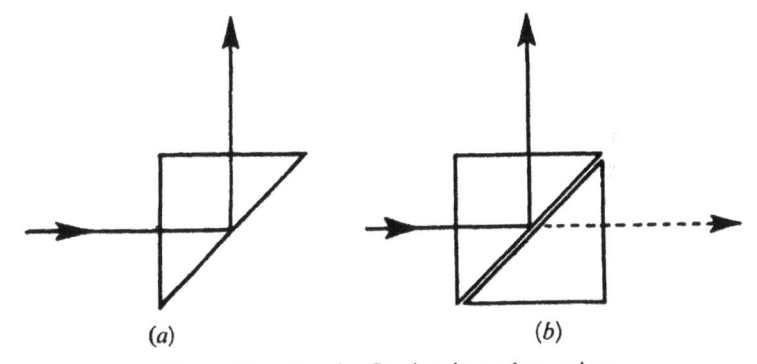

Figure 22.—Total reflection in a glass prism.

back and forth within the nucleus, of a wave which is reflected back and forth from the walls of the 'potential container', and we can even imagine it as a stationary wave. But now we must explain how such a wave can be reflected from these walls at all. Obviously, this phenomenon is based on a process for which we can find an analogy in the total reflection of light at the surface separating two transparent refractive substances, such as, for instance, the boundary plane between glass and air. This phenomenon occurs, for instance, in the prisms of a Zeiss field glass. If light falls perpendicularly on one of the short sides of a right-angled triangular glass prism, it will strike the surface of its hypotenuse at an angle of 45 deg. However, according to the law of refraction, it cannot emerge from this surface by refraction, but will be totally reflected in the same way as it would be by a perfectly reflecting mirror. (Figure 22 (a).)

If a second prism is placed near the first one (Figure 22 (b)), no change will take place so long as the distance between them is sufficiently great. But as soon as this distance becomes very small, a little light can pass into the second prism, too. For in total reflection, a small quantity of light energy always seeps through the reflecting surface, but only for a very short distance, of the order of magnitude of the wavelength. If the second prism is brought sufficiently close, the light which has seeped through can penetrate into the second prism and can travel further in the ordinary manner. The closer the two surfaces are to each other, the greater is the quantity of light that goes through, and if the surfaces are pressed firmly together, no more total reflection occurs at all.

Something quite analogous to this phenomenon takes place in the case of the matter waves of the alpha particles. In this analogy, the interior is to be regarded as the equivalent of one prism, and the outside space as the equivalent of the other one, while the potential barrier between them is the equivalent of the layer of air separating the two prisms. Some waves will always get through the barrier, and the thinner is this barrier, the larger proportion of the waves will escape into the outer space. In this connection, 'barrier' means that part of the potential curve which rises above the horizontal energy level of the particle. From this it follows automatically that the higher is this level—and hence, the greater is the energy of the alpha particle—the more transparent will be the barrier to the waves of the alpha particle, for the potential barrier that must be overcome, will be proportionately narrower. Thus, assuming that originally the waves were present in the interior only, as more and more time elapses, we will find an increasingly larger portion of them outside, too, and will find that this portion increases with the increase of the energy of the particle.

Now this description in terms of the wave aspect must be translated back into the language of the particle aspect. In doing so, we must bear in mind that in the course of our discussion of the conditions prevailing within the extranuclear atomic structure we stated that the local density of the matter waves constitutes a measure of the probability of encountering alpha particles at the point under consideration. But this density outside of the nucleus, and hence the probability of finding alpha particles outside of it, increases rapidly when we deal with a particle which is rich in energy, and more slowly when our particle is less rich in energy. The more energy a particle has, the greater is the probability that it will no longer be found inside, but outside of the nucleus—in other words, that it is hurled out within a very short time. This is where we find the explanation of the fact that in the case of alpha particles rich in energy the decay probability is very much greater than in the case of those less rich in energy. The exact mathematical expression of this idea leads to a very satisfactory agreement with the Geiger-Nuttall law.

If we wish to give a quite summary description, in terms of the particle aspect, of the effect just discussed, we may state that contrary to all expectations based on the law of the conservation of energy, after a certain period of time (the length of which is a matter of chance) the particle is able to break through the potential barrier as though through a tunnel. It is customary therefore to speak of a *tunnel effect*.

We have already mentioned that although all the elements above zinc in the periodic system might be expected to be prone to alpha decay, this decay can, nevertheless, be observed in the heaviest elements only. It would be logical to assume that no sharp dividing line existed between the well-known alpha emitters and those elements which would seem capable, from the point of view of energetics, of emitting alpha radiation, but are never observed to do so. It may well be possible that a major number of the elements above zinc in the periodic table do actually emit alpha particles, but both the energy and range of these particles are very small. Both these factors would account amply for our practical inability to observe them. However, such radioactivity would be imperceptible mainly because the low magnitude of energy corresponds to an extremely small decay probability. For even though such an element might emit an alpha particle at infrequent intervals, it is nevertheless stable for all practical purposes, even when measured by cosmical time standards.

II. THE BETA EMITTERS

Let us now turn our attention to the second type of spontaneous nuclear transmutation, which takes place with beta radiation—the emission of electrons or positrons, accompanied by a neutrino. Such transmutations take place when they are compatible with the conservation laws, and thus in particular when energy is liberated from the nucleus in the process. Here, too, there arises the question, why a nucleus in which all the prerequisites of a transmutation are given, does not change immediately.

The differences in the lifetimes of the various beta emitters are smaller than those among the alpha emitters. The half-lives of beta emitters vary from a few seconds to a few years. Only very few of them have a much longer half-life.

We cannot account for beta radiation by the same arguments which we used to explain the properties of alpha emitters; at any rate, these arguments are not applicable to the electron emitters, for every electron carries a negative charge, and consequently, when outside of the nucleus, it is attracted by the latter. Therefore, in this case, the potential barrier is nonexistent. However, since there is no essential difference between electron emitters and positron emitters, we cannot suppose that there is an analogy between positron emitters and alpha emitters either. Furthermore, there is the fact that the electrons and positrons, together with their accompanying neutrinos, do not constitute integral parts of the nucleus as do the alpha particles, but are created from the nuclear field only at the instant when the nuclear transmutation actually occurs. An analogy with the emission of photons from the extranuclear atomic structure seems to be a more logical one.

Empirically, it is generally true that the decay probability of the beta emitters also increases with the energy of the beta particle. It must be borne in mind that every electron and positron is accompanied by a neutrino, and as already pointed out, the reaction or decay energy is shared by this pair according to statistical laws. Therefore, the energy of the fastest beta particle, the neutrino of which happened not to receive any share of the decay energy, is decisive for the decay energy.

Now, in order to understand, at least qualitatively, the relationship between energy and decay probability, let us borrow from the theory of electric waves. For a beta radiation should actually be likened to the emission of light from the extranuclear atomic structure. We are thinking now in terms of the wave aspect, and therefore will speak of an electron or a positron, plus a neutrino, as a wave emanating from the atomic nucleus. We will liken this wave to the electric wave sent out by a radio aerial.

Planck's law, $E = h\nu$, holds good for these matter waves, too; it correlates their energy, E, with their frequency, ν. In this case, E is the energy of the beta particle, or more exactly, that portion of the decay energy which falls to the share of the electron. If, then, the decay probability increases as the energy, the higher is the magnitude ν, the greater will be the decay probability, and therefore the shorter the wavelength of the radiation. If the dipole moment of a vibrating aerial is kept constant which can be achieved by maintaining the peak voltage between the condenser plates of the oscillatory circuit constant, then the radiation will be all the stronger the higher the frequency in the circuit. The intensity of the waves at a point is proportional to the fourth power of the frequency. The situation with respect to the matter waves of beta radiation is quite analogous, with the sole difference that their energy is proportional not to the fourth power of the frequency, but—as shown by a closer theoretical analysis by Fermi—to its sixth power. It follows, therefore, that the decay probability (the energy emitted per second divided by the energy of the individual decay) is proportional to the fifth power of the energy available for the decay process.

The above considerations represent the actual conditions in their general outlines only, but not quantitatively correctly. Before we can gain a complete understanding of the subject, we must discuss still another consideration. We have assumed the dipole moment of the aerial to be constant. However, this cannot be expected to be the case for every one of the various nuclear types. On the contrary, considerable individual differences must be expected to exist, and do in fact exist. One would therefore expect the decay probability to be the product of two factors, the first being determined by the dipole moment (and as the latter is determined quantitatively by the size of the nucleus in question, it may vary widely from one nucleus to another), and the second one being proportional to the fifth power of the decay energy. Figure 23 shows a summary of the lifetimes, and hence of the decay probabilities and energies, of the beta emitters. Whenever a power of a quantity is supposed to be proportional to a power of another quantity, it is most convenient to record on the co-ordinate axes not these powers themselves but their logarithms, for the logarithms must be linearly related to each other. Thus, Figure 23 shows *log T*, the logarithm of the half-life, *T*, measured in seconds, as a function of *log E*; the scales on the two axes differ by the factor 5. If the fifth-power law is valid, all the beta emitters must lie on a straight line forming an angle of 135 deg. with the axes. Actually, the empirical points lie, mostly, between two such straight lines; the distance between these two straight

THE NUCLEAR REACTIONS

lines provides a measure of the fluctuation of the dipole moment from one nucleus to another.

Before proceeding to the discussion of other decay processes, let us write down the formulae of a few more processes

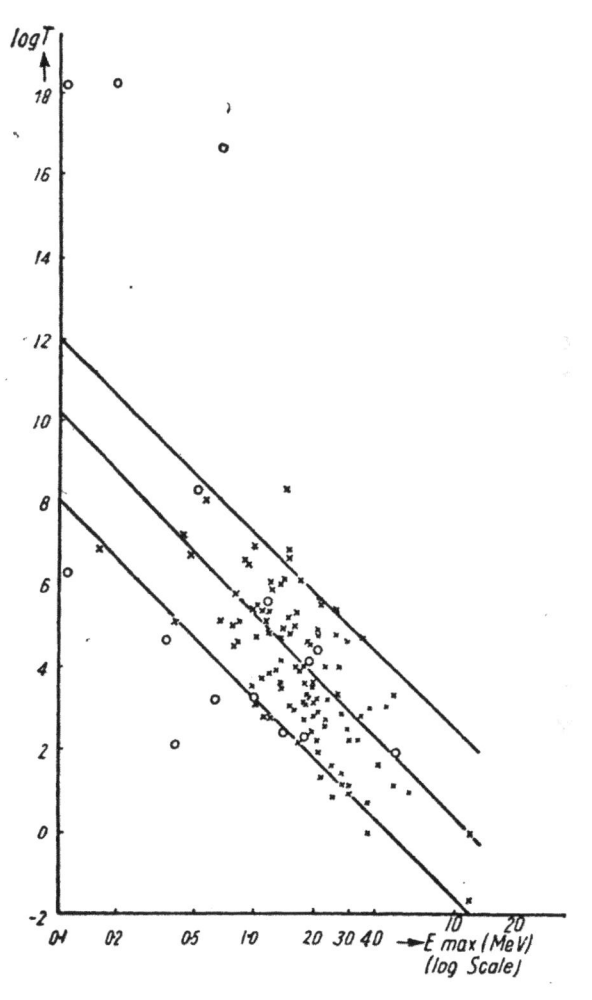

Figure 23.—Life and energy of natural and artificially produced beta radiators (*Sargent diagram*).

involving both alpha and beta radiation. When the uranium atom $_{92}U^{238}$ decays under emission of an alpha particle, $_2He^4$, the product of the process is an atom of mass number 234 and atomic number 90—uranium X_1, $_{90}UX_1^{234}$. We write this process as follows:

$$_{92}U^{238} \rightarrow {}_{90}UX_1^{234} + {}_2He^4$$

As always, the mass numbers and atomic numbers (superscripts and subscripts) must balance on both sides of the arrow.

The boron atom $_5B^{12}$, for instance, emits an electron and changes into the carbon atom $_6C^{12}$. Simultaneously with the emission of the electron, a neutrino is emitted, of which both the rest mass and the charge are 0; its symbol, therefore, is $_0\nu^0$. We write, consequently:

$$_5B^{12} \rightarrow {}_6C^{12} + {}_{-1}e^0 + {}_0\nu^0$$

An example of a positron emitter is the nitrogen atom $_7N^{13}$, which changes into the carbon atom $_6C^{13}$ by the emission of a positron, $_1e^0$ and a neutrino, i.e.:

$$_7N^{13} \rightarrow {}_6C^{13} + {}_1e^0 + {}_0\nu^0$$

III. OTHER TYPES OF SPONTANEOUS NUCLEAR TRANSMUTATION

Alpha and beta radiation are by far the most frequent types of radioactive transmutation. However, there is also a third process, already mentioned in a cursory manner, which is more or less the converse process of electron emission. In cases where it is advantageous, from the point of view of energetics, for a proton in the nucleus to change into a neutron, this transformation can be accomplished in two different ways. We have already discussed one of them—the emission of a positron. However, the same result is attained when the nucleus absorbs one of its own planetary electrons, in which case the electron, naturally, vanishes (as an absorbed photon vanishes) since its charge offsets the charge of the proton and changes the latter into a neutron. This process is called *electron capture*. As in this

THE NUCLEAR REACTIONS 127

process no charged particle is emitted, the process can only be observed, from outside, by the consequent x-ray emission, *K-radiation*, since the absorbed electron usually comes from the innermost electron shell, the *K*-shell. As this process reduces the nuclear charge (atomic number) by one unit, the number of remaining electrons (the original number of planetary electrons minus one) is sufficient for the newly formed atom. But the remaining planetary electrons must now re-group themselves; an electron originally occupying a position in shell farther from the nucleus falls into and occupies the place vacated by the captured electron. This produces the x-ray emission, which is the *K*-radiation characteristic of the element in question. At the same time, in order to conserve the angular momentum or spin value of the atom—since the captured electron had a spin value of $1/2\ h$—a neutrino is emitted.

Such a process occurs, for instance, in the beryllium nucleus $_4Be^7$, an artificially produced radioactive isotope of beryllium. Capturing a planetary electron, it changes into the lithium nucleus $_3Li^7$. The formula expressing this process is:

$$_4Be^7 + _{-1}e^0 \rightarrow {_3Li^7} + {_0\nu^0}$$

Such processes are not so very rare. They are frequently associated with fairly long half-lives. The half-life of $_4Be^7$ is approximately fifty-three days. In our Table IV, the atomic nuclei which change by *K*-capture are indicated by circles.

And finally, there is still another process, discovered by Hahn and Strassmann in 1938. This is the reaction already mentioned in which nuclei split into two parts of approximately equal size. Occasionally it may occur spontaneously, too. But this process will be discussed under the heading of artificially induced nuclear transmutations.

IV. ARTIFICIALLY INDUCED NUCLEAR TRANSMUTATIONS

As was first demonstrated by Rutherford, a nuclear transmutation can be produced artificially by shooting some particle

NUCLEAR PHYSICS

into the nucleus. In the majority of the cases, this particle will remain in the nucleus without producing any reaction. But the result may be that a particle of some kind is emitted in turn by the nucleus. If this emitted particle is not of the same kind as the first particle, this reaction means a nuclear transmutation.

Bohr formulated the following theory about such a reaction: When a particle is hurled at the nucleus and actually strikes it, it usually remains in it, for the simple reason that it is held fast there by the exceedingly powerful nuclear forces. As a result, the energy of this particle becomes distributed very rapidly

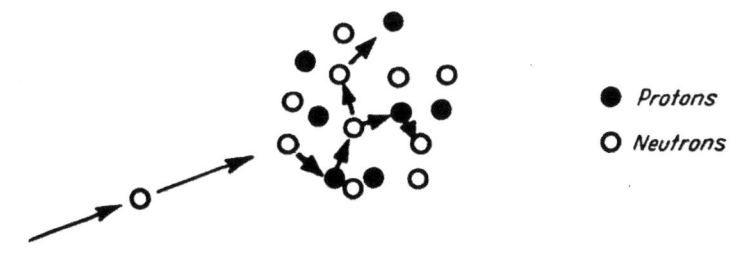

Figure 24.—Neutron penetrating a nucleus.

among the other particles in the nucleus, and eventually throughout the entire nucleus. Figure 24 shows a schematic picture of a nucleus. A neutron is approaching it from the outside. The white and black circles represent the nuclear neutrons and protons. As shown by the arrows, these nuclear particles receive an impact from the foreign neutron and, in turn, collide with other particles. When the neutron has penetrated into the interior of the nucleus and its energy has become distributed among all the particles within it, the situation can be expressed in the simplest form as follows: The atomic nucleus is being heated. A pile of sand becomes heated in exactly the same manner when a bullet is fired into it. This reaction is analogous to the heating of a microscopic structure, if we bear in mind that an increase in the kinetic energy of the molecules in the sand pile corresponds to an increase in its temperature. In a nucleus, instead of molecules we deal with neutrons and protons, and the kinetic energy is associated with a certain temperature of the nucleus.

The temperature attained by a nucleus in this way—for instance, when a particle with an energy of approximately 8 Mev. is shot into it—has, in conformity with the laws of the kinetic theory of gases, the order of magnitude of ten thousand million degrees. It is roughly a thousand times higher than the highest temperatures that otherwise occur in the universe, the temperature in the interiors of the fixed stars. However, in this case, these high temperatures only affect the infinitesimally minute territory of that one nucleus.

If we imagine the nucleus in this state as a highly heated drop of liquid, the logically inevitable conclusion is that the nucleus will evaporate in consequence of the high temperature. This means that after a short while, some particle or another will emerge from the nucleus—generally that particle, the emergence of which is most advantageous from the point of view of energetics, in other words, the one the emergence of which requires the least energy. The energy necessary here corresponds to the heat of evaporation in an actual liquid. The nucleus cools off as a result of having yielded up this energy. Occasionally, a second particle will emerge, too. Or else, the residue of the energy, which is not sufficient for the emission of another particle, departs in the form of a gamma ray, as a photon. In a certain respect, this process is like the case of an incandescent drop of liquid which, due to its high temperature, emits visible light, too.

The process just described can continue unimpeded only if the particle used to bombard the nucleus is uncharged, i.e. if it is a neutron or a photon. But if the bombarding particle is a proton or an alpha particle, it must come up in the vicinity of the nucleus, against the potential barrier, the height of which is in proportion to the magnitude of the charge of the bombarded nucleus. The 'projectile' would be slowed up in its flight against the nucleus, and in many cases—especially those involving heavy nuclei—it would either come to rest at some distance from the nucleus or would be deflected so far from its path that it would never strike the nucleus. Therefore, in order actually to hit a heavier nucleus, the charged particle would have to be accelerated by means of exceedingly high voltages, which can be produced only in apparatus especially designed for this purpose. For this reason, nuclear transmutation by means of charged particles is possible, as a rule, in relatively light atoms only.

On the other hand, there is no potential barrier for uncharged particles, and therefore they can be used to transmute atoms of any mass number. Transmutations of heavier atomic nuclei by photons, i.e. gamma rays, have been observed by Bothe and his collaborators, while the transmutation by neutrons was first demonstrated by Fermi. The neutron will become incorporated in the nucleus, in many cases at any rate, and the surplus energy then will be carried off by one or more gamma-ray photons.

While this reaction signifies a change in the nucleus, it does not involve any change in the chemical properties of the atom. The chemical properties change only when the resulting nucleus is unstable because it has too many neutrons within it. In that case, a subsequent process takes place, in which a neutron changes into a proton, by the emission of an electron, with the result that the nucleus becomes a nucleus of the element whose atomic number is one higher.

Since a neutron can approach a nucleus unimpeded, the velocity plays no essential role in the process, unlike the case of charged particles. On the contrary, slow neutrons are frequently more effective than fast ones, for they remain longer in the vicinity of the nucleus, and so the probability of their eventual capture by the nuclear forces is greater than for fast neutrons.

Experiments have shown that this probability can be extraordinarily great for neutrons having a certain, but not too high, energy. According to Bethe, this becomes quite logical when we consider the neutron under the wave aspect, i.e. as a wave incident on the nucleus. The nucleus is a structure capable of vibration, and as such, it can enter into resonance with any wave it encounters, if the frequency of that wave coincides with one of its own fundamental frequencies. In this case, an exceedingly strong selective absorption of the wave occurs, a process quite familiar to us in connection with the absorption of light. Now the frequency of the wave is a function of the velocity of the neutron, and therefore there exists a very definite velocity where the requirements for resonance are fulfilled and the wave is especially strongly absorbed by the nucleus. But translated back again into terms of the particle aspect, this means that the probability of the capture of the neutron by the nucleus is a very high one. This dependence on the velocity is often described in terms of a *nuclear cross-section*. Let us assume that nuclei are spheres, neutrons are points, and no force of any kind is acting between them at all. In that case, the larger are the cross-sections of the spheres, the greater would be the probability of striking a nucleus with neutrons shot at it at random. In this model, the nuclei would apparently have cross-sections of different sizes for neutrons of different velocities. Under particularly favourable conditions, this nuclear cross-section may be about 10,000 times larger than the actual geometrical cross-sectional area of the nucleus. This means an exceedingly high probability of capture. This is why Fermi, who discovered this circumstance, first applied the device already mentioned of slowing down the neutrons (which are always rather fast when they are first produced) to thermal velocity. He made them pass through a hydrogen-containing substances, such as water or paraffin. Hydrogen is the best substance for this purpose, because the mass of a proton is approximately the same as that of a neutron, and also because the conditions for a rapid exchange of energy are most favourable here, in conformity with the laws of elastic impacts. When such slow neutrons are captured by a nucleus, they are accelerated within the range of attraction of the nuclear forces sufficiently to produce that heating of the nucleus which has already been discussed.

Conversely, too, it is naturally easier for a neutron to emerge out of such a heated nucleus than it would be for a charged particle. In fact, the charged particle has to 'climb over' the potential barrier in order to get out, whereas for the neutron the potential barrier is non-existent. This is the reason why nuclear transmutations involving the emission of a proton or an alpha particle are relatively rare in heavy nuclei where the potential barrier is high.

Of course, when a nucleus is heated to a very high degree, a larger number of charged and uncharged particles can come out of it, analogously to the evaporation of a liquid drop. With laboratory apparatus it is very difficult to impart such high energies to the nucleus. But particles possessing the vast energy of 1,000 Mev., or even more, are present in cosmic radiation. If such a particle happens to hit a nucleus, the result of the encounter is that the nucleus is heated to such a high degree and emits numerous protons and neutrons, at times even heavier fragments, such as helium or lithium nuclei. When a nuclear disintegration of this kind actually takes place in the emulsion on a photographic plate, the nuclear particles leave tracks which become visible when the plate is developed. This technique was developed by Blau, Wambacher and Schopper, and has recently been improved considerably by Powell and his collaborators. Figure 25A shows such a photograph. It is a much enlarged reproduction; the range of the protons in the emulsion is actually less than 1 mm. We are thus looking here at an actual record of the disintegration of an atom, in which at least forty particles or so have 'evaporated' out of the nucleus. For, in addition to the charged particles, many neutrons, too, are sure to have been liberated. Figure 25B shows a similar, still more

THE NUCLEAR REACTIONS

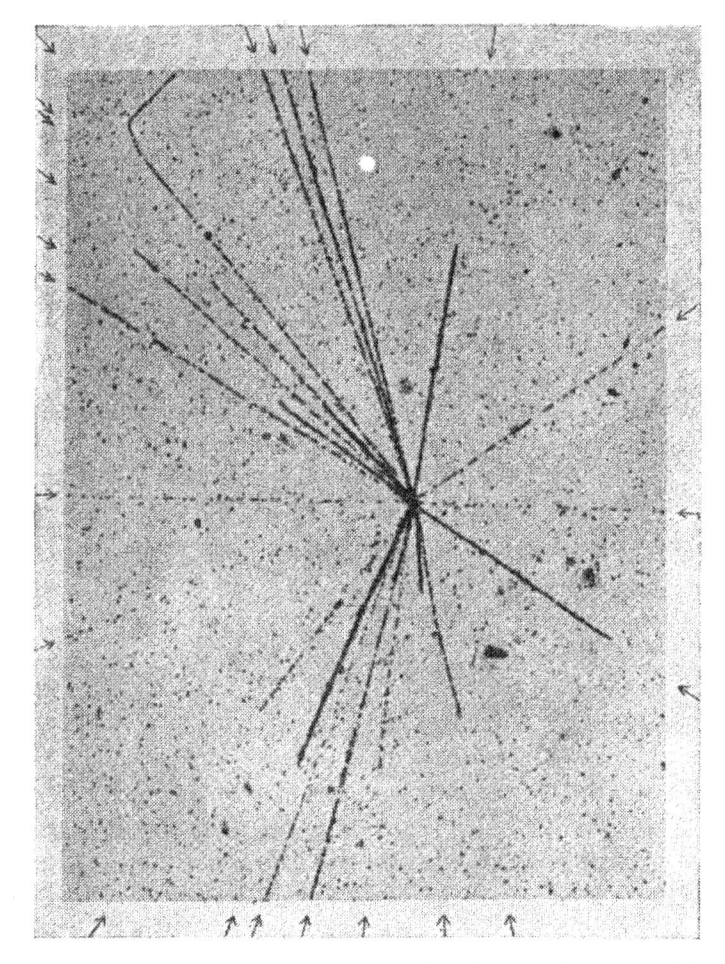

Figure 25A.—Disintegration of a nucleus by cosmic ray particle of very big energy (*after Powell and Occhialini*).

complicated process. In the disintegration of one nucleus a particle has been liberated, which subsequently induces a transmutation reaction in a second nucleus. The difference in the density of the silver grains in the individual tracks is due to the difference in the velocities of the nuclear particles. The

134   NUCLEAR PHYSICS

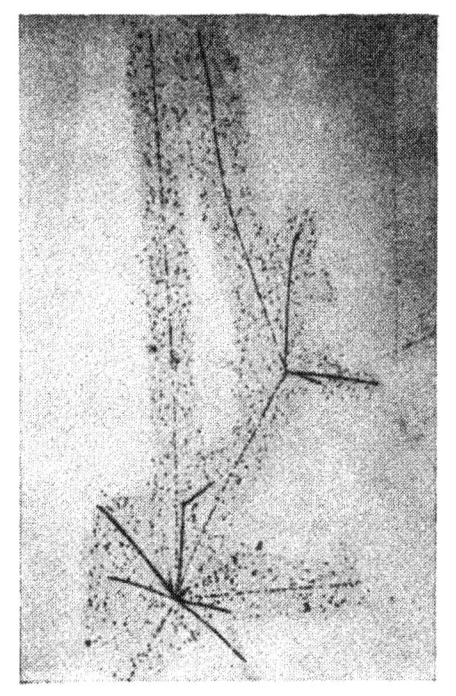

Figure 25b.—Primary and secondary nuclear disintegration.

faster they are, the more thinly they are spread, and the longer is the track which they leave. Of course, actually, the particles are dispersed in all directions, and therefore, the majority of their paths appear more or less foreshortened.

A last, quite especially important case of the disintegration of nuclei is *nuclear fission* (the splitting of nuclei), discovered by Hahn and Strassmann in Berlin in 1938. This is what occurs in this case: It may happen that a heated nucleus ejects no individual particles at first, but begins to vibrate as a whole, utilizing all or part of the introduced energy for the excitation of these vibrations. This is often the case when a uranium nucleus is bombarded by a neutron. A reaction ensues then, which is shown diagrammatically in Figure 26. The nucleus, orginally spherical in shape, will vibrate first so as to assume alternately an elongated and a flattened elliptical form. The longitudinal deformation may reach a degree where the nucleus—like an iron rod which is about to break in two—becomes very thin approximately in its middle, and finally breaks into two more or less equal parts. Splinters fly off, as a rule, since several neutrons are ejected.

The fact that such fission is possible at all and is most likely to occur in the heaviest nuclei, is easy to understand. The nuclear forces which guarantee the stability of a nucleus

Figure 26.—Nuclear fission.

are opposed by the electric force of repulsion which increases with the nuclear mass, since, considered on the whole, the charge also increases with the mass. The repulsion produces a decrease of the binding energy per particle, and thus a decrease of the stability with the increase in the mass. If in addition to this factor, the stability of the nucleus is imperilled by its vibration, the electric force of repulsion is able to make its effect felt more freely. Above a certain amplitude of vibration, it is able to increase still further the vibration initiated, and eventually to tear the nucleus apart.

This fission of the nucleus may take place in various ways. As a rule, the two fragments are not of equal size. For instance, when the rare uranium isotope $_{92}U^{235}$ has absorbed a neutron, it may split into a strontium atom, $_{38}Sr^{90}$, and a xenon atom, $_{54}Xe^{144}$, and two neutrons, i.e.:

$$_{92}U^{235} + _{0}n^{1} \rightarrow _{38}Sr^{90} + _{54}Xe^{144} + _{0}n^{1} + _{0}n^{1}$$

In this case, too, naturally, the totals of the mass numbers must be equal on both sides of the arrow, and the charge numbers must balance likewise.

However, instead of producing the results thus indicated, the nuclear fission of this uranium atom may produce from the latter, for instance, the strontium atom $_{38}Sr^{88}$ and the xenon atom $_{54}Xe^{146}$, plus two neutrons, or atoms of two different elements and a different number of neutrons. In fact, in such fissions the formation of a very considerable number of different elements has been observed. The precise way in which the uranium nucleus splits (in which fission occurs) is a matter of chance to a certain extent.

We have thus obtained a general bird's eye view of the different possibilities of nuclear transmutation. Neutrons can be used to transmute all nuclei; charged particles are most suitable for producing a transmutation of the lighter nuclear species. Among the charged particles which primarily enter into consideration are protons and deuterons, which are capable of being sufficiently accelerated in electric fields, as well as the alpha particles, both natural and artificially produced by an intensive acceleration of helium nuclei in electric fields. Finally, there are several nuclei which can be heated to such a high degree by a sufficiently energy-rich gamma-ray photon that they emit a neutron; this emission changes their mass only, but not the chemical properties of that particular atom.

In conclusion, let us discuss a few examples which have played an especially important part in the progress of nuclear physics. Let us take these in their chronological order.

In 1919, Rutherford accomplished the first artificial nuclear transmutation, changing nitrogen atoms into oxygen atoms by bombardment by alpha particles. A proton (being a hydrogen nucleus, it is written $_1H^1$) was liberated in this process, which can be expressed as follows:

$$_7N^{14} + {_2He^4} \rightarrow {_8O^{17}} + {_1H^1}$$

The oxygen atom thus formed is the rare oxygen isotope of

THE NUCLEAR REACTIONS 137

mass number 17. Figure 10 shows such a process in the cloud chamber.

Another important transmutation reaction led to the discovery of the neutron by Joliot-Curie and Chadwick, in 1932. A beryllium nucleus, $_4Be^9$, was bombarded by an alpha particle, and a carbon nucleus, $_6C^{12}$, plus a neutron was obtained, thus:

$$_4Be^9 + {}_2He^4 \to {}_6C^{12} + {}_0n^1$$

As the third of these important reactions, let us mention the first transmutation of nuclei by artificially accelerated particles,

Figure 27.—Transmutation of lithium nucleus into two helium nuclei by a proton (*after Kirchner*).

accomplished by Cockroft and Walton, using protons, in the same year, 1932. The protons were accelerated by a high-tension apparatus of 600,000 volts. When such a proton hits a lithium nucleus, $_3Li^7$, the following reaction occurs:

$$_3Li^7 + {}_1H^1 \to {}_2He^4 + {}_2He^4$$

In other words, two alpha particles, or helium nuclei, are formed. Figure 27 (after a photograph by Kirchner) shows this

138   NUCLEAR PHYSICS

process taking place in the cloud chamber. The end of the discharge tube, in which the protons are accelerated, is visible. From there they impinge on a piece of metallic lithium. We can see the tracks of two alpha particles start there and proceed in opposite directions. (The other visible track does not belong

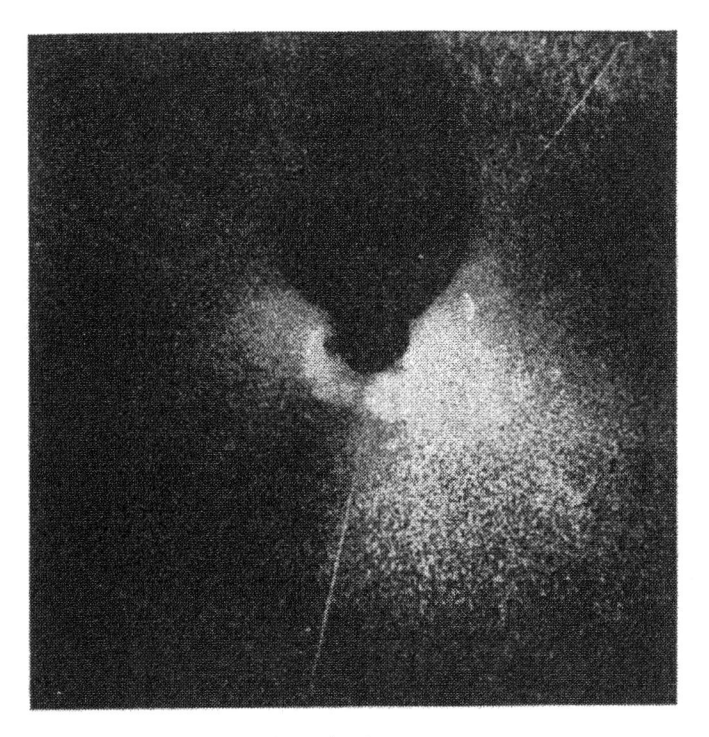

Figure 28.—Transmutation of a boron nucleus into three helium nuclei by a proton.

to this reaction.) A similar reaction is shown in Figure 28. Here a boron nucleus changes, in consequence of the absorption of a fast proton, into three helium nuclei, i.e.:

$$_5B^{11} + {_1H^1} \rightarrow {_2He^4} + {_2He^4} + {_2He^4}$$

(Here, too, an alpha particle is visible, which does not happen to belong to this reaction.)

Finally, in view of the ancient history of these problems, let us ask whether we are able today to make gold out of mercury, which used to be the dream of the alchemists of old, and in what way this might be possible. To answer this question, it is sufficient to take a look at Table IV. It shows that mercury and gold are immediate neighbours, so that just one single step would be necessary to accomplish the transmutation of one into the other. In other words, by pure chance, the alchemists of past centuries happened to be on the right track when they endeavoured to turn mercury into gold. According to our present knowledge, mercury has 7 stable isotopes, with mass numbers ranging from 196 to 204, while gold has only one, with the mass number 197; this mass number is absent among the known isotopes of mercury. If the mercury isotope of mass number 196 is irradiated by neutrons, so that a neutron is incorporated in a mercury nucleus, the otherwise unknown nucleus of mass number 197 must be the result. This nucleus must be unstable, or else it would have been observed already. It will therefore change, by positron emission or K-radiation, into the stable gold nucleus of the same mass number. Thus the following two reactions occur, in this order:

(1) $\quad _{80}Hg^{196} + {}_{0}n^{1} \rightarrow {}_{80}Hg^{197}$

(2) $\quad _{80}Hg^{197} \rightarrow {}_{79}Au^{197} + {}_{1}e^{0}$

The mercury nucleus $_{80}Hg^{196}$ has changed into the gold nucleus $_{79}Au^{197}$.

Thus, fundamentally, the nuclear physicist would have no difficulty in producing gold out of mercury. However, this transmutation has never yet been actually recorded.

One might wonder why this transmutation has never been performed as yet. But the reason is that the profit would be far too small. Unfortunately, the mercury isotope $_{80}Hg^{196}$ is extremely rare; it represents not more than about 0·1 per cent. of the natural isotopic mixture of mercury. If mercury is bombarded by neutrons, only one neutron in a thousand will happen to become incorporated in one of these nuclei. From the less rare isotopes we cannot produce gold, but either another mercury isotope or thallium. It is, of course, conceivable that our goal could be reached more easily by a bombardment by fast neutrons. The mercury isotope $_{80}Hg^{198}$ is about 100 times more frequent than $_{80}Hg^{196}$. If we succeeded in heating this mercury nucleus of mass number 198, by neutron bombardment, to such a high degree that it would emit two neutrons simultaneously, we would get the gold nucleus $_{79}Au^{197}$, viz.:

(1) $\quad _{80}Hg^{198} + {_0}n^1 \longrightarrow {_{80}}Hg^{197} + {_0}n^1 + {_0}n^1$

(2) $\quad _{80}Hg^{197} \longrightarrow {_{79}}Au^{197} + {_1}e^0$

But in order to nip false hopes in the bud, let it be stated here that this method would indubitably be many million times more costly than the customary methods of getting gold.

7. THE TOOLS OF NUCLEAR PHYSICS

I. THE METHODS OF DETECTION AND OBSERVATION

The following sections of this book will deal with the tools and methods available to the nuclear physicist both for generating and observing the phenomena discussed in the preceding lectures. These procedures call for immense quantities of energy, and the most powerful instruments known to technical science must be used to supply them. Yet, the material results achieved even with these vast stores of energy are extremely small. Consequently, for these studies it is indispensable to have extraordinarily sensitive instruments, for the phenomena which are to be studied take place in one individual atom or, at best, in a very few atoms—in structures which are inconceivably small according to ordinary conceptions.

We shall begin with the instruments of detection and study. The oldest method is the *scintillation method.* When a very fast particle—an alpha particle for instance—impinges on a zinc sulphide screen, a reaction occurs there which produces a weak flash of lightning, a *scintillation.* It is therefore possible to observe the impact of each individual particle of sub-atomic order of magnitude, as the impact of bullets on a plastered wall can be observed, and we can also count the particles this way. It is, however, a poor policy to depend on the unaided eye, which is bound to grow tired little by little during the process of counting. This method is scarcely ever employed these days. It has been taken up recently by recording the weak flashes not with the eye but by electric amplification.

The *ionization chamber* supplies the fundamental principle on which most of the modern methods of observation are based. Let us attempt to describe this apparatus in a very rudimentary form: A gold-leaf electroscope consists of an earthed metal box containing an insulated metal rod with two gold leaves which spread apart when an electric charge reaches the rod. (Figure 29.) An inverted metal hood, charged by a battery to about 100 volts relatively to the earth, is placed above the electroscope and insulated from the latter. When a charged particle—an alpha or beta particle, or even a gamma-ray photon—enters the space between the electroscope and the hood, it tears electrons off the air molecules. These electrons attach themselves to other molecules and form negative ions, while the molecules thus deprived of electrons remain behind as positive ions. As shown in Figure 29, such ions are created all along the track of the particle. If the hood is positively charged, the positive ions

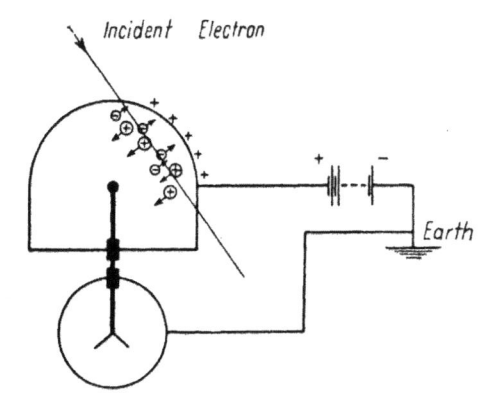

Figure 29.—The principle of the ionisation chamber.

stream to the rod of the electroscope, and the negative ions to the hood. Thus the electroscope becomes charged, and its leaves spread apart. Of course, the apparatus in this rudimentary form is not sensitive enough to detect individual particles. Indeed a far more delicate apparatus, such as a string electrometer, while able to detect a rather weak radiation, cannot register individual particles. The method of detecting small charges has been improved considerably in various ways and respects, according to the particular nuclear-physical purpose in view.

The first apparatus of general applicability (for the detection of individual particles) was *Geiger's point counter* (Figure 30). Fundamentally it is simply a vastly improved ionization chamber. In this apparatus, a metal rod is drawn to a fine point, and the chamber is given a fairly high voltage, with the result that a strong electric field is created in the vicinity of the point. When a charged particle or a gamma-ray photon flies past and liberates electrons there, these liberated electrons are so strongly accelerated by the intense field that they, in turn, are able to tear electrons off air molecules. These electrons, in turn, are able to do the same thing, and thus the number of electrons

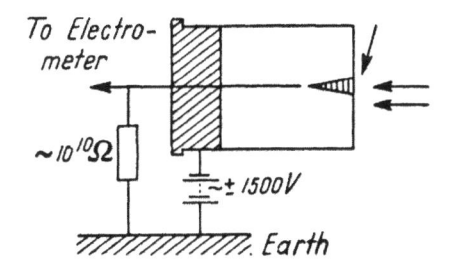

Figure 30.—Point counter (*after Geiger*).

liberated increases like an avalanche, and their number ceases to increase only where the field is weaker. But during this process, such enormous numbers of them are produced even by the effect of one single particle or photon that they can be detected by the means available to us.

Under certain conditions, when the voltage is not too high, the multiplication of the number of electrons always increases by the same factor. Therefore we speak of a *proportional region* of the counter and also of a *proportional count*. The factor just mentioned may be of the order of magnitude of 1,000 or even higher. But if the voltage on the counter is increased beyond a certain limit, we pass beyond the proportional region. In that case, the electrons liberated by the particle start a genuine glow discharge, so that the result is a ten-millionfold or even a hundred-millionfold increase. In that case the discharge must be stopped, so that the counter may once again be ready for a new particle. In this *resolving region*, the amplification is independent of the number of the primarily liberated electrons. The amplification continues always until the very moment when a glow discharge begins to take place.

About fifteen years ago, this point counter was considerably improved by Geiger and Müller, and has become the counter which is still by far the most important observing apparatus

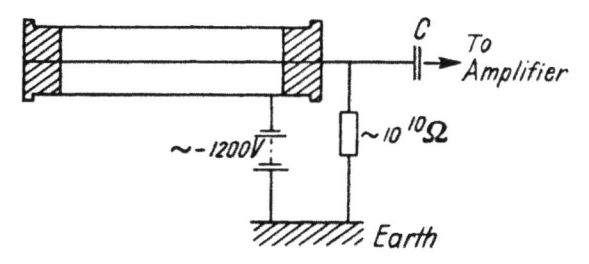

Figure 31.—Geiger-Müller counter.

of the nuclear physicist. In principle it is very similar to the original Geiger point counter, except that instead of a point, a thin wire is placed in its centre. (Figure 31.) Usually, it is filled not with air, but with a mixture of argon, under a pressure of 60 to 80 mm. of mercury, and alcohol vapour under a pressure of about 10 mm. of mercury. But there are also several variants. The wire is earthed through a very big resistance, and the outer cover has a potential difference of 1,000–1,200 volts relative to the earth.

The situation here is similar to that in the point counter. At a lesser voltage, a proportional amplification, by the factor 1,000, takes place. When the voltage is higher, a glow discharge begins; the device is now operating in the resolving region. From the moment of the inception of the glow discharge, the wire, which may be connected to a condenser, becomes strongly

THE TOOLS OF NUCLEAR PHYSICS

charged, as does also the condenser, since owing to the very high resistance, an appreciable time must elapse before the charge flows to the earth. Therefore, during this period, both the wire and the condenser, C, connected with the wire, are at a certain voltage which can be amplified by methods commonly employed in broadcasting technique. As is customarily done in most of these measurements, a counting device, similar to a telephone counter, can be attached, or the voltage can be transferred to a loudspeaker, and thus one can count and register every particle that passes through the counting tube.

The number of the ions produced by beta and gamma radiation is small. Therefore, big amplification is required, and it is customary to work in the resolving region. Furthermore, since the beta particles are not very penetrating, thin-walled tubes are used for counting beta particles; on the other hand, for gamma-ray photons thick-walled tubes are used, so as to keep other types of radiation out as much as possible. When dealing with alpha rays, which produce far more electrons, big amplification can be dispensed with, and it is possible to work in the proportional region. The advantage of this method is that, if the counting device is properly connected, it does not react to other types of radiation. The latter produce weak potential impulses only. A special amplifier, called the *thyratron*, which transmits impulses above a certain wavelength only, permits the weaker impulses to be eliminated, so that only those produced by alpha radiation are counted. This is important mainly because, in addition to the radiation under investigation, all other conceivable types of penetrating radiation are roaming through space. In the first place, electrons are liberated everywhere, even in the counter itself, by cosmic radiation which cannot be screened sufficiently by any known means. Secondly, no existing substance is entirely free from radioactive impurities, so that even the material of which the counter is made tends to release impulses occasionally. Such limited effects are simply just inevitable with these measuring devices. When counting

146 NUCLEAR PHYSICS

alpha particles, the counting tube must be equipped with a thin window of mica, to enable the particles to enter, since they would not be able to pass through anything thicker.

Another very important instrument of the nuclear physicist is Wilson's *cloud chamber*, the operation of which was explained in our second lecture. The advantage of this device consists principally in the fact that it permits us to obtain a visual record of nuclear processes, thus showing simultaneously a great many of the details of the process.

Figure 32 shows a simple sketch of the cloud chamber. The upper section contains air, saturated with water vapour. It is

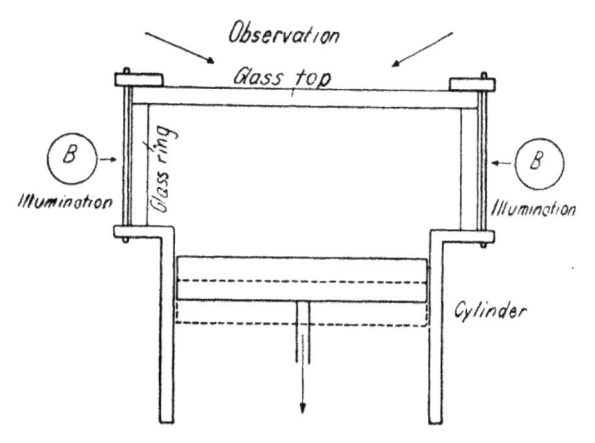

Figure 32.—A simplified sketch of Wilson's cloud chamber.

covered with a glass plate at the top to permit observation; below there is a movable piston, covered by a damp layer of gelatine, so that the air above it is kept saturated with water vapour. The light necessary for observing the cloud tracks is admitted through an aperture in the side. The piston is suddenly moved downward, so that the air expands adiabatically and cools. The result is that the water vapour becomes supersaturated and the ionization produced by a particle entering the

THE TOOLS OF NUCLEAR PHYSICS 147

chamber causes condensation along its path—the well-known cloud tracks.

Many phenomena which one would like to observe in the cloud chamber, in particular the phenomena of cosmic radiation, are extremely rare, and the observer may have to wait for several hours before they eventually take place. The chances of encountering just such a reaction when observing an expansion, are very small indeed. Therefore, if we had to rely on chance alone for the eventual observation of such rare occurrences, studies of this kind would require a very great deal of time. However, this handicap is eliminated by skilfully connecting the cloud chamber with a counter which acts, so to speak, as a sentry at the gate of the cloud chamber. The counter is adjusted so that it will react to the specific phenomenon which it is desired to observe in the cloud chamber. If such a phenomenon actually takes place, the counter immediately effects the expansion through an amplifier. This takes place so fast that the ions formed in the chamber have not yet diffused away from the paths of the particles, and they are therefore visible as cloud tracks. This is the method which has yielded the most important data known concerning the nature of cosmic radiation, during the past decade.

Finally, the photographic plate, too, can be used as a detector of charged particles. An example has been shown already in Figure 25.

These methods enable us to count or detect every radiation with which an electric charge is associated (alpha and beta radiation, as well as all other types of nuclear debris which carry an electric charge), and also gamma-ray photons, so that the question of the detection of neutrons remains the only one still to be explained. Since neutrons carry no electric charge, they do not themselves produce ionization, and therefore we must rely on the observation of a secondary effect in order to detect their presence. The simplest device for this purpose is a boron counter. The inside of the wall of this counter is lined with boron or some boron compound, and the tube is used in the proportional region, so that it will count alpha particles only. As the neutrons strike the boron layer, they produce there the following nuclear reaction:

$$_5B^{10} + {}_0n^1 \rightarrow {}_3Li^7 + {}_2He^4$$

This reaction produces fast helium nuclei, i.e., artificial alpha particles—one alpha particle per neutron. Every neutron evoking a nuclear reaction causes the counter to react with an impulse, in which the lithium nucleus will also participate. Not every neutron, by any means, striking the counter causes a nuclear reaction; many of them traverse the tube without any effect whatsoever. Nevertheless, the counter registers a number of neutrons proportional to the actual total number of neutrons. The constant factor of proportionality being as yet unknown.

Another frequently employed method consists in placing a *tracer* at the point where neutrons are suspected to be present. A tracer is a substance—for example, a piece of silver foil—made artificially radioactive by a neutron-induced nuclear reaction. The following two reactions will then take place in the silver, in this order:

(1) $\quad _{47}Ag^{107} + {}_0n^1 \rightarrow {}_{47}Ag^{108}$
(2) $\quad _{47}Ag^{108} \rightarrow {}_{48}Cd^{108} + {}_{-1}e^0$

In other words, the silver isotope of mass number 107 first changes into another silver isotope, which has the mass number 108. The latter is unstable, has a half-life of 22 seconds, and emits an electron and thus changes into a nuclear isobar, the cadmium atom $_{48}Cd^{108}$. The $_{47}Ag^{108}$ atom must be unstable, because its nucleus has 61 neutrons and 47 protons, in other words, it is a 'doubly odd' nucleus.

Since, as we have already seen, slow neutrons are usually more prone to capture by a nucleus than are fast neutrons, a boron counter will register a greater number of slow neutrons than fast ones. If a counter of this type is brought into the vicinity of a source of fast neutrons, a loudspeaker can be used, which is capable of making the individual impulses audible at a certain average rate, say, one per second. As we already know, neutrons can be slowed down by being made to pass through a hydrogen-containing substance, such as paraffin. If the counter is surrounded by paraffin, the impulses will multiply very considerably and a crackling sound will be audible. Thus, contrary to the naïve assumption that the application of paraffin is bound to reduce the effect, a quite considerable intensification of the latter is the result. This method for the increase of the output in nuclear transmutations by means of slowing down neutrons, is employed very frequently in nuclear physics.

II. THE PROCEDURES FOR PRODUCING NUCLEAR TRANSMUTATION

As a general rule, particles very rich in energy are required to produce a nuclear transmutation. It is only when the transmutation is induced by neutrons that the energy content of the bombarding particle is frequently reduced deliberately as much as possible. But neutrons, to start with, must be produced by a nuclear reaction induced by fast particles, such, for instance, as the bombardment of beryllium by alpha particles.

Nature itself provides us with the most convenient source of energy-rich particles: The natural alpha emitters. To be sure, the radiation of even the most powerful radioactive preparations is always relatively weak, and sufficient for the transmutation of only a small number of atoms. On the other hand, in addition to the alpha particles, still other types of particles, namely, fast protons and deuterons, are needed in order to produce all possible sorts of nuclear transformation.

The most logical method of producing particles of high energy consists in a very strong acceleration of charged particles by a very high voltage, possibly 1,000,000 volts or more. Of course, direct-current voltage must be used. It is of course much more difficult to produce such a high direct-current voltage than it would be to generate an alternating-current voltage of the same magnitude.

The system known as the *Greinacher circuit* (Figure 33) is one very frequently employed today in high-voltage generators. Two columns, with condensers, C, are connected by a system of valves, V, each of which permits a beam of electrons to pass through in one direction only, for instance, in the direction indicated by arrows in our diagram. The entire system is designed so that the point d, for instance, can be charged positively (but not negatively) relative to the earthed point a, without being discharged. Similarly, c can be charged positively relative to d, likewise f relative to c, e relative to f, etc. Now, if an alternating-current voltage (usually between 200 and 300 kv or so) is applied, by means of a transformer, T, between the two columns, the points d, c, f, etc., become positively charged through the valves until the alternating-current voltage of, for instance, the point d never falls, throughout an entire period, below that of a, which may be taken as equal to 0; for otherwise a current would still flow through the valve V_1. Thus, if the peak voltage of the transformer is $+E$, the potential of the point d, in a stationary state, fluctuates between 0 and 2E, and the point c has the constant potential 2E. No current flows then through the valves. Similarly, we find that in a stationary state the points e, g, i have the constant potentials 4E, 6E and 8E, while the potential of the points f, h, k fluctuates between 2E and 4E, between 4E and 6E, and between 6E and 8E, respectively. When, for instance, a current enters at the point i, the potential there decreases slightly, and the valves let through a beam of electrons in the direction of the arrow; these electrons carry the charge along, so that the potential at the point i cannot drop very far below 8E. Thus, proceeding by n steps, we obtain a 2n-fold of the peak voltage of the transformer—for instance, when the initial voltage E is 200 kv and three steps are used, the direct-current voltage ultimately obtained is 1,200,000 volts.

THE TOOLS OF NUCLEAR PHYSICS 151

Figure 34 shows an exterior view of the high-voltage generator of the *Kaiser Wilhelm Institut für Physik* in Berlin-Dahlem. The slanting parts are the valves, and the globes correspond to the points c, e, f in Figure 33.

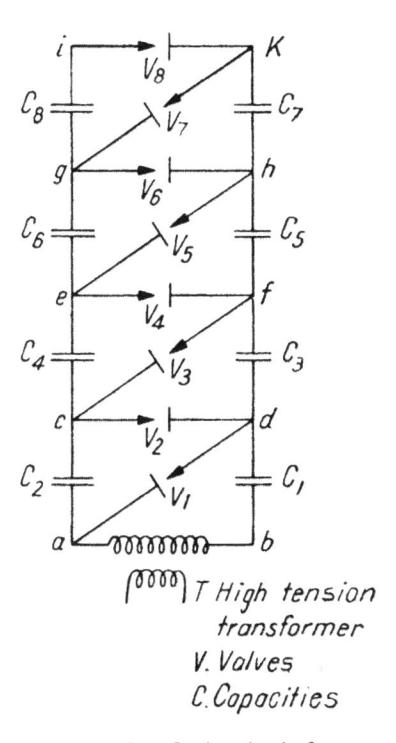

Figure 33.—Greinacher's form.

The high voltage thus produced must now be used to accelerate charged particles. These latter originate as canal rays in an ordinary discharge tube. They then enter the highly evacuated accelerator tube, with the high voltage drop between its terminals. At the end of the accelerator tube they hit the substance which is meant to be transmuted.

The disadvantage of this apparatus is that it is extremely expensive. Therefore efforts have been made to achieve the

L

152 NUCLEAR PHYSICS

same results by simpler means. In this connection the *Van de Graaff high-voltage generator* deserves attention. It is based on the old, and now hardly ever used principle of the influence machine. This generator consists of a large hollow metal sphere (Figure 35), or cylinder, serving as conductor, with a pulley within it and another pulley underneath it. A wide closed belt

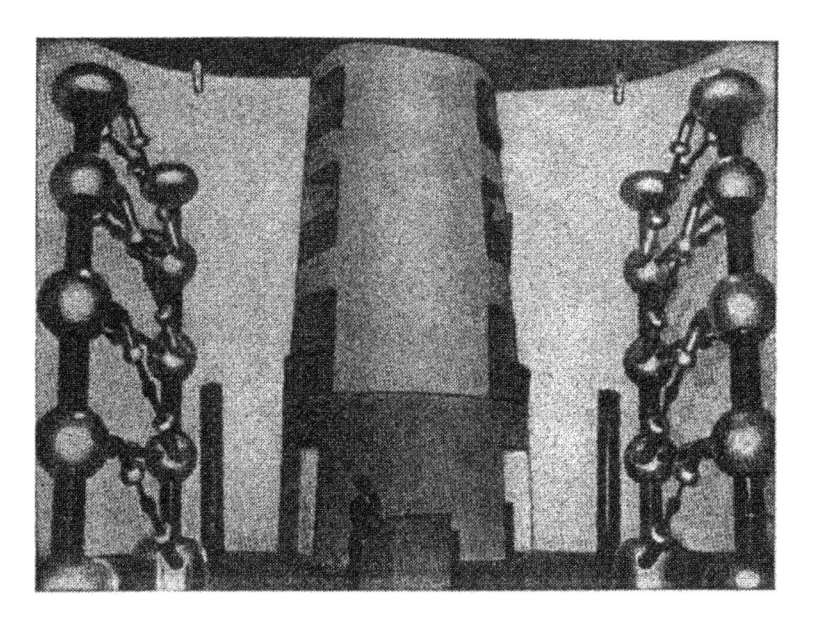

Figure 34.—High tension generator of the Kaiser Wilhelm Institute (Max Planck Institute) in Berlin-Dahlem.

of some insulating material, for instance, silk, travels on these two pulleys. Outside of the conductor, electric charge is sprayed on the belt by means of a rectifier and a corona comb, and the belt, carrying this charge, enters the conductor, in which the charge is removed by a second corona comb and is transferred to the conductor. The conductor can thus be charged up to any desired voltage. Certain limitations are imposed by the dimensions of the space in which the generator is installed, since

THE TOOLS OF NUCLEAR PHYSICS 153

when a certain voltage is reached (the magnitude of which depends on the dimensions of this space and of the conductor itself) a spark jumps across to the walls and discharges the conductor. In 1939, there existed as yet no operating generator capable of attaining more than 2,000,000 volts. Figure 36 shows the largest installation of this nature, which was under construction in the United States several years ago. It is designed to

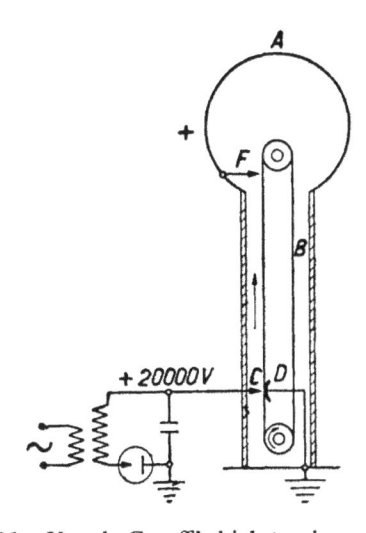

Figure 35.—Van de Graaff's high tension generator.

produce 5,000,000 volts, and therefore built in a very large space, an old airship hangar. It has two conductors, supposed to be charged with opposite charges, in order to produce the double voltage for the discharge tube.

By far the most efficient apparatus for the production of fast particles is the *cyclotron*, invented by the American Lawrence. It is based on a very interesting principle, that of very frequently repeated acceleration by the same, not very high, voltage; thus it has, among other advantages, the good feature that it dispenses with the high voltages which are so difficult both to obtain and to control. The essential part of the cyclotron is a very large electromagnet which creates a powerful, very homogeneous and wide magnetic field of 10,000–15,000 oersteds between its pole pieces. The pole pieces are placed quite near to each other and the space between them is well evacuated. When a moving particle enters such a field, it describes a circular

Figure 36.—Van de Graaff's high tension apparatus.

path, the radius of which is proportional to the velocity of the particle (Figure 37). Therefore, the velocity of the particles is proportional also to the circumference of the circle, and as a result, particles of the same kind, even though having different velocities, require exactly the same time to complete a full revolution. The space between the pole pieces houses two semi-

THE TOOLS OF NUCLEAR PHYSICS 155

cylindrical boxes, called *dees*, insulated from each other; between these dees, a potential difference of 30–100 kv is produced by a high-frequency generator. The result is a high-frequency alternating field in the small space between the dees. The frequency of this alternating field is regulated so as to correspond exactly to the period of the revolution of the particles in the

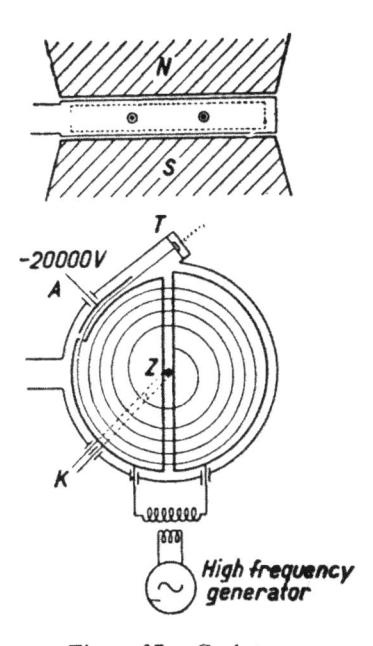

Figure 37.—Cyclotron.

magnetic field. The charged particles are made to enter the space between the pole pieces, near the centre (Z). There they come under the influence of the electric field; they attain a certain velocity and move in a semicircle in the space inside a dee where there is a magnetic field only. Proceeding in this manner, they reach the channel between the dees at the exact moment when the electric potential drop there is exactly equal, but opposite in direction, to what it was at the moment of their initial acceleration. They are now of course moving from one dee to the other in the opposite sense, which is that of the electric field, and in consequence are further accelerated. So the same process is repeated over and over again, and the velocity of the particles continues to mount. They move in an approximately spiral orbit, composed of semicircles, always further and further outward, until they are hurled through a window (T), designed so as to be penetrable by them, to perform their appointed task, the production of nuclear reactions.

The adjustment of such an apparatus calls for a great deal of technical skill. Moreover, the cyclotron is a machine of such dimensions as are seldom encountered in any other appliance used in physical research. Let us illustrate this by mentioning a few figures. The pole pieces of a cyclotron in use in the United States for some time are 95 cm. in diameter. The magnet of this cyclotron produces a magnetic field of 14,000 oersteds, and 60 tons of iron and 10 tons of copper went into its construction. The production of the magnetic field of 14,000 oersteds requires an input of 30 kilowatts. If this cyclotron is used to accelerate deuterons, these will emerge from it with an energy of 9 Mev.—in other words, as if they had passed through a potential drop of 9,000,000 volts. A current of only 0·1 milliampere flowing through this drop of potential would represent a power of nearly 1 kilowatt (900 watts to be precise). As each particle carries an elementary quantum of electricity, $1·6 \times 10^{-19}$ coulombs, it is easy to compute that this current is the equivalent of roughly 6×10^{14} particles per second.

Figure 38 shows an exterior view of such a cyclotron. The windings of the magnet are visible; between the pole pieces of the magnet are the dees, in which the particles begin their acceleration. The beam emerging from the cyclotron is also visible in the photograph.

Numerous cyclotrons are already in use in the United States. Several have been built in Europe, too. Germany has had one

Figure 38.—Cyclotron.

Figure 39.—Magnet of the giant cyclotron.

since 1944, in the *Kaiser Wilhelm Institut* in Heidelberg, destined primarily for medical use. The universal recognition of the importance of the cyclotron is demonstrated by the amount of money spent on it in the United States, where the rough structure of a giant cyclotron was completed in 1940; its size makes it look more like a battleship than a scientific instrument. Its pole pieces are 4·7 metres in diameter, and its magnet is 17·8 metres in length (Figure 39). Its foundation contains 1,200 tons of concrete; the magnet contains 3,700 tons of iron and 300 tons of copper, wound in a strip of 10·2 cm. in width and 6 mm. in thickness. The frame of the magnet is made of 36 steel plates, each 5·5 mm. in thickness. The magnetic field intensity is 10,000 oersteds, and the frequency of the alternating electric field corresponds to a wavelength of 39 metres. Lawrence completed this cyclotron after the war, and it has enabled him to accelerate deuterons up to 100 Mev. and alpha particles up to 200 Mev.

In other words, the cyclotron is an extremely costly and also extremely complex apparatus. However, it is still by far the most useful nuclear-physical research instrument designed for the same purpose. In the United States it has made it possible to accomplish many nuclear reactions which would not have been feasible by any other means.

8. THE PRACTICAL APPLICATIONS OF NUCLEAR PHYSICS

I. THE EXPLOITATION OF ATOMIC ENERGY FOR USEFUL PURPOSES

In studying the practical applications of nuclear physics, it is helpful to start out with analogies with chemistry. Chemistry deals with the combination of different elements in more complex substances, the chemical compounds, or conversely, with the separation of elements from such compounds, whereas nuclear physics deals with the transmutation of one element into another. Chemical processes serve two fundamentally different purposes: First, they can be used to convert less valuable materials into others of greater value, as, for instance, the combination of carbon and hydrogen to form benzol; secondly, a chemical conversion can be utilized in order to obtain energy, as, for instance, the combustion of coal to form carbon dioxide, in order to produce heat. It is obvious that these two applications are not independent of each other. It is often the case that a material is produced solely in order to be used as a source of energy—as, for instance, benzol.

The same is true, we may say, of nuclear processes. They can be applied, first, to produce a more valuable material out of one of lesser value, and secondly, they can be instrumental in producing energy.

In order to gain an idea of the order of magnitude of the energy which can be obtained through nuclear reactions, let us make use of another analogy with chemistry. The combustion of carbon with oxygen resulting in the formation of carbon dioxide can be written in the form of an equation which also shows the energy yielded by this process:

$$C + O_2 \longrightarrow CO_2 + 96 \text{ kilocalories}$$

This formula relates to units of 1 mole and 1 gramme-atom, and states that the combustion of 1 gramme-atom = 12 grammes of carbon with 1 mole = 32 grammes of gaseous oxygen produces 1 mole = 44 grammes of carbon dioxide and liberates 96 kilocalories of heat.

Another example is the combustion of oxygen with hydrogen, resulting in water. The formula describing this process, likewise relating to units of 1 mole, is:

$$H_2 + \tfrac{1}{2}O_2 \longrightarrow H_2O + 68 \cdot 4 \text{ kilocalories}$$

This means that the quantity of heat liberated is 68·4 kilocalories for every mole of water formed.

It may be said, quite generally, that these *heat emissions* of chemical processes are, throughout, of the order of magnitude of approximately 100 kilocalories per mole.

Now let us write the formula of a nuclear reaction, which happens to be one frequently applied in nuclear physics today, and which according to Döpel, can be released by ordinary canal rays having an energy corresponding to 5–10 kilovolts, viz.: the reciprocal reaction of two deuterons, which results in the formation of a hydrogen nucleus of mass number 1 and a hydrogen nucleus of mass number 3. The formula of the reaction, again relating to 1 mole, is:

$$_1D^2 + {_1D^2} \longrightarrow {_1H^1} + {_1H^3} + 100{,}000{,}000 \text{ kilocalories}$$

In other words, the two deuterons do combine at first, but the compound nucleus thus formed splits up immediately into the two nuclei indicated above, since this is more favourable from the point of view of energetics, as is demonstrated by the tremendous heat emission of 100,000,000 kilocalories per mole. The heat emissions in the majority of the other nuclear reactions are of a similar order of magnitude. We see, therefore, how important nuclear reactions become as soon as they can be produced on a really large scale. The energy yield of nuclear reactions is about one million times greater than that obtainable by chemical processes. In other words, the nuclear 'combustion' of matter yields a millionfold of the energy released by a chemical combustion of the same quantity of the substance.

In nature, such nuclear reactions occur, as a rule, on a small scale only—due to the effect of cosmic radiation and of the radiation of radioactive substances. The fact that the energy released by these natural phenomena is not observed, is due to the fact that the quantities of energy produced are far too small, and also far too scattered.

Nevertheless, it is not quite right to claim that the energy liberated by nuclear reactions has never before played any part in nature. On the contrary, it may be claimed with full justification that in the long run, we owe our entire existence on earth to such processes. For, firstly radioactivity plays an extremely important part in determining the temperature and climate of the earth's surface, and secondly, it is owing to such reactions in the sun that it shines on the earth and sustains life on our planet.

Nuclear reactions on a large scale are in fact taking place in the interiors of the stars. We know today that the energy which the stars—including our own sun—radiate into space is the product of such nuclear reactions. The source of this radiated energy was a puzzle for a long time, the solution of which called for a great deal of mental skill. We know that our sun has shone upon the earth with approximately the same intensity for at least two thousand million years, and in former years nobody could comprehend why it had not expended its total energy a long time ago. The solution of this problem had to wait for nuclear physics, and today we are in the position to specify exactly the only process that can possibly account for it. This solution was established as a result of three studies carried out by Atkinson and Houtermans, v. Weizsäcker, and Bethe. We cannot describe here the way that led to this solution; we can merely quote the result. It is a question of a certain sequence of formulae, which may be stated here, as follows:

(1) $_6C^{12} + {}_1H^1 \longrightarrow {}_7N^{13}$

(2) $_7N^{13} \longrightarrow {}_6C^{13} + {}_1e^0$

(3) $_6C^{13} + {}_1H^1 \longrightarrow {}_7N^{14}$

(4) $_7N^{14} + {}_1H^1 \longrightarrow {}_8O^{15}$

(5) $_8O^{15} \longrightarrow {}_7N^{15} + {}_1e^0$

(6) $_7N^{15} + {}_1H^1 \longrightarrow {}_6C^{12} + {}_2He^4$

This series consists partly of proton-induced reactions ($_1H^1$ is the symbol of the proton), partly of transmutations by the emission of a positron ($_1e^0$). The initial substances are carbon of mass number 12, and hydrogen. It is a known fact that the stars consist to a large extent of hydrogen, and that carbon also is present in them in small quantities. The hydrogen is already in the form of protons, since as a result of the enormous velocities due to the high temperatures (10–20 million degrees C.) prevalent in the interiors of stars, practically all the hydrogen atoms have been stripped of their single planetary electrons. This enormous velocity enables the protons to penetrate the other nuclei.

So, first, a nitrogen nucleus is formed from an ordinary carbon nucleus and one proton (1). This nitrogen nucleus is unstable and changes by the emission of a positron, into a carbon nucleus (2). This newly produced carbon nucleus is a heavy isotope of that which started the process. (The entire process is well known from laboratory experiments, too.) Now, another bombardment by a proton changes this carbon nucleus into an ordinary nitrogen nucleus (3). The latter changes, by the absorption of another proton, into an unstable oxygen nucleus (4), which immediately emits a positron and thus becomes a heavy isotope of the first nitrogen nucleus (5). Finally, the entire process is concluded by another absorption of a proton, and the emission of a helium nucleus (in other words, an alpha particle) results once again in an ordinary carbon nucleus.

A 'balance sheet' of this series of reactions shows the following picture: First, there is a carbon nucleus, $_6C^{12}$, which captures four protons, step by step. At the end, there remains the same $_6C^{12}$ nucleus, with a helium nucleus, $_2He^4$, plus the two positrons emitted in reactions No. 2 and No. 5. The summary of this 'balance sheet' can be put as follows: One helium nucleus and two positrons have been formed from four protons. The net result of the process, therefore, is expressed by the following summary formula:

$$4\,(_1H^1) \longrightarrow {}_2He^4 + 2\,(_1e^0)$$

The helium nucleus consists of two protons and two neutrons; hence, its charge is two units smaller than the charge of the four protons. This difference is accounted for by the two positrons. Two of the four protons have thus changed into neutrons.

Since the masses of the protons and of the helium nuclei are exactly known, the energy balance of the entire process can be determined on the basis of the last formula. The energy liberated in such a process is 25·5 Mev., or—computed for moles and converted into kilocalories—600,000,000 kilocalories per mole. This is six times as much as in the process mentioned above.

That which takes place here can be expressed as follows: In the interiors of the stars, hydrogen is converted into helium by nuclear 'combustion', and this process liberates the vast energies continuously radiated by the sun and stars. It has been said occasionally, in jest, the sun is 'heated by coal'. But this is not quite correct. The coal—carbon—acts here merely as a catalyst, and is not consumed in the reaction.

This example ought to be sufficient to show that immense stores of energy are liberated in nuclear reactions, if these take place in sufficient quantities of material. Let us further add that there is actually good reason to believe that older stars are poorer in hydrogen than younger ones, and that this fact points to a gradual consumption of hydrogen.

Now why is it that we have been unable to produce similar quantities of energy in the laboratory before the discovery of the fission of uranium, assuming for the purposes of this question that we had at our disposal a very powerful source of neutrons, consisting of about 100 grammes of radium mixed with beryllium? Compared with the sources normally available in the laboratory, this would be a very plentiful one, indeed. If we were to use this source of radiation to irradiate ordinary table salt (sodium chloride) for a whole day, an estimated 20,000 million chlorine atoms would be changed into radioactive sulphur atoms. This is a very high number, and the sulphur produced would be actually very strongly radioactive. But unfortunately, this quantity of sulphur is extremely small—just about one thousandth of a millionth of a milligram. The energy obtained from it is correspondingly small in quantity—just six-millionths of a kilocalorie.

However, the largest cyclotron in operation at the present time can increase the neutron intensity to a point approximately a thousand times what used to be regarded as the upper limit. This intensity equals roughly that of a neutron source consisting of 100 kilograms of radium mixed with beryllium. In this case, both the quantity of the substance and the energy have been multiplied by approximately 1,000, and yet they remain extraordinarily small. At any rate, the energy yield is still just a minute fraction of the energy input required by the cyclotron in order to achieve this result. Therefore, a practical exploitation of nuclear energy will be possible only when the transmutation reactions take place spontaneously—and when such a reaction induces a second one and that one induces still another, and so forth, so that finally, as in chemical reactions, most of the nuclei have become transmuted in a *chain reaction.*

II. URANIUM FISSION AND CHAIN REACTION

In 1938, Hahn and Strassman discovered the process of fission in the uranium nucleus, which process has already been described on pages 134 and 135. A uranium nucleus is hit by a neutron, splits into two approximately equal parts, and in most instances, several neutrons are hurled forth. This reaction constitutes the foundation of modern nucleonics, since it permits the occurrence of the chain reaction, mentioned at the end of the preceding paragraph.

Now a few more details of this process of splitting, or fission, of nuclei. It releases a great deal of energy, in other words, the two fragments of the nucleus are hurled apart with enormous velocity. This energy—about 150 Mev. per fission—can be determined empirically by measuring the velocity of the fragments, or computed from the formula of mass defects (see page 72). Since this is a typically *exothermic* process (a process involving a yielding up of energy), it can take place also without an actual bombardment of the nucleus by neutrons. But the spontaneous fission of nuclei is such a rare phenomenon that technically it has no importance whatever.

Nuclear bombardment by neutrons is capable of producing nuclear fission more or less easily in the different elements at the end of the periodic table (Table III). Certain nuclei can be split even by slow neutrons (i.e. neutrons of thermal velocities). Every one of these is a nucleus with an odd mass number, above all the uranium nucleus $_{92}U^{235}$ and the nucleus of the element plutonium, which will be discussed later. In these nuclei, the small amount of heating linked with the capture of a neutron is sufficient to produce fission. On the other hand, there are other nuclei which can be split by neutrons having a high velocity only. Thus, for instance, neutrons of at least 1 Mev. are required to produce a fission of the uranium nucleus $_{92}U^{238}$. As American experiments with the huge California cyclotron have shown in the course of the past few years, if neutrons (or other nuclear 'projectiles') of more than 30 Mev. are used, it is possible to produce fission in the nuclei at the end of the periodic system as far down as tin.

When nuclei of the latter type are bombarded by slow neutrons, the energy of which is not sufficient to cause fission, the neutrons will be either reflected from the nucleus (mostly with a loss in velocity) or captured (see page 90). Thus, for instance, the nucleus $_{92}U^{238}$ may change by the capture of a neutron into the nucleus $_{92}U^{239}$, which then changes spontaneously, by beta emission, into $_{93}Np^{239}$ (*neptunium*) and $_{94}Pu^{239}$ (*plutonium*). The names 'neptunium' and 'plutonium' were suggested for elements 93 and 94 by the American scientists who first studied the properties of these nuclei. The processes just mentioned can be written as follows:

$$_{92}U^{238} + _{0}n^1 \longrightarrow _{92}U^{239} \longrightarrow _{93}Np^{239} + _{-1}e^0$$
$$_{93}Np^{239} \longrightarrow _{94}Pu^{239} + _{-1}e^0$$

The plutonium nucleus, $_{94}Pu^{239}$, thus formed, emits an alpha particle and changes into uranium, $_{92}U^{235}$, although this is a rarer reaction. The half-life of this plutonium nucleus is about 24,000 years. As already explained, the capture of neutrons, as by the $_{92}U^{238}$ nucleus, is a reaction the occurrence of which is more likely in the case of certain values of the neutron energy, where the incident neutron wave vibrates in resonance with the vibrations of the nucleus.

In the fission process, as a rule, a few neutrons are also hurled out from the nucleus; thus, for instance, in the fission of the $_{92}U^{235}$ nucleus, two or three neutrons are emitted. This phenomenon, first verified as a fact by Joliot in 1939, makes possible the occurrence of the chain reaction necessary for the technical exploitation of nuclear energy. When a fission reaction takes place in a sufficiently large mass of—for instance— pure $_{92}U^{235}$, the neutrons released by it collide with other $_{92}U^{235}$ nuclei, causing fission in the latter, too, with the result that still more neutrons are liberated; these, in turn, produce fission in still other nuclei, and so on, so that finally, the entire mass of the substance is changed and a vast quantity of energy is released in the process. Owing to the high velocity of the neutrons, the whole process takes place in less than one-millionth of a second. It is, therefore, quite obvious that a sufficient quantity of pure $_{92}U^{235}$ (or a sufficient quantity of pure $_{94}Pu^{239}$) is an explosive of inconceivable power. Atomic bombs are made of these substances—and their destructive power is well known. A prerequisite for the occurrence of the chain reaction is, of course, that the mass of the 'explosive' should be sufficiently large, for if it is too small, the neutrons would escape through the surface before causing the fission of other nuclei. Therefore, a small mass or quantity of the above-mentioned 'explosives' is totally harmless. However, as soon as their mass or quantity exceeds a certain magnitude, the explosion occurs immediately, and spontaneously. Therefore, the atomic explosion is started by combining small pieces of the 'explosive' into a big piece (by throwing them together), and the big piece thus formed explodes at once.

Oppenheimer was the scientific leader of the atomic bomb construction in America. The production of sizable quantities of the 'explosives' $_{92}U^{235}$ and $_{94}Pu^{239}$, calls for such a tremendous technical effort that only the vast industrial capacity of the United States of America was in the position to afford it. In Germany, their production was not attempted during the war, for the capacity of the already overburdened German industry would not have been sufficient for it.*

III. THE URANIUM REACTOR

The use of nuclear energy for peaceful purposes is more important than the manufacture of explosives. In order to harness this energy for peaceful purposes, it is essential to be able to produce a chain reaction which can be kept under control, so as to permit the withdrawal of just the right quantity of energy

* For details, see the report at the end of this book (p. 189).

required at any given moment. Fortunately, it is possible to produce such a chain reaction in natural uranium, which is a mixture of two uranium isotopes—$_{92}U^{238}$ and $_{92}U^{235}$, in the proportion of 140 to 1, and it is not necessary to enrich the natural uranium in the rare $_{92}U^{235}$ first.

However, a chain reaction does not take place in pure metallic uranium, for the neutrons emitted in the fission reaction collide far more frequently with $_{92}U^{238}$ nuclei than with $_{92}U^{235}$ nuclei. They are reflected by the $_{92}U^{238}$ nuclei, as a rule, without loss in velocity, and are eventually captured by one of these nuclei with which they resonate, so that they are lost for the chain reaction. But pieces of uranium can be encased in a *moderator*, a substance that slows down the neutrons. This permits the neutrons to be carried quickly past the resonance values and their speed to be reduced to thermal velocities. But despite the relatively small number of $_{92}U^{235}$ nuclei, the thermal neutrons are more easily captured by them than by the $_{92}U^{238}$ nuclei and, as a rule, they then produce fission in the $_{92}U^{235}$ nuclei. The effect of the moderator thus is that the neutrons are seldom captured by $_{92}U^{238}$ nuclei. If the moderator selected is a substance which absorbs thermal neutrons to a very small extent only, and if the apparatus is of a sufficient size to prevent the escape of too many neutrons through the surface before entering into reaction with the $_{92}U^{235}$ nuclei, a chain reaction can be started. Practically, the only substances suitable for use as moderators are heavy water (D_2O) and absolutely pure graphite, the absorption coefficient of both of which is very small for thermal neutrons. In a uranium *reactor* (as an apparatus consisting of uranium and a moderator is called) built of pieces of uranium and heavy water, the following chain reaction takes place: A neutron, released by fission, leaves the piece of uranium —possibly after a few collisions with uranium atoms—and reaches the heavy water. There, due to collisions with deuterons, it loses velocity, and wanders around in the moderator at thermal velocity, until eventually it happens to collide again

PRACTICAL APPLICATIONS

with a piece of uranium. It then produces a new fission in a $_{92}U^{235}$ nucleus, as a result of which again two or three neutrons are liberated—and so it goes on. While the chain reaction in the atomic bomb is provoked by fast neutrons, the velocity of which is reduced by inelastic collisions to only a little below their original speed, the chain reaction in the uranium reactor is propagated by slow neutrons.

This chain reaction can be easily controlled and guided. The uranium is heated by the disintegration of the uranium atoms. The result is a spreading of the points of resonance in the $_{92}U^{238}$, and so a greater number of neutrons are captured by these nuclei. The heating thus atomatically brakes down the chain reaction, so that the entire apparatus becomes stabilized at a certain temperature, the magnitude of which depends on its size and geometric design. Furthermore, it is possible to introduce, into the reactor, from outside, some substance that absorbs neutrons (cadmium is the most suitable material for this purpose) and consequently will act as an added moderator of the chain reaction and regulate the temperature artificially. Once a certain temperature is reached, the reactor will maintain it quite independently of the quantity of energy removed from it. If a considerable amount of energy has been removed— in consequence for instance of good thermal conductivity—the reactor cools off a little at once. The frequency of the disintegration processes immediately increases enormously, and the original temperature is re-established.

Figure 40A shows the interior of the model of a uranium reactor, composed of uranium and heavy water. It was installed, during the war, in a cellar carved in natural rock, in the village of Haigerloch in Württemberg, by a working team of the *Kaiser Wilhelm Institut* (*Wirtz, Bopp, Fischer, Jensen, Ritter*). The photograph shows a great many disks of metallic uranium, suspended by chains from a lid which can be lowered into a tank containing heavy water. The tank itself is enclosed in a thick layer of graphite, which, however, is hardly visible

170 NUCLEAR PHYSICS

in this picture. Figure 40B is a schematic digram of the ur reactor; the shaded area is the graphite coating. The apparatus stood in a large water tank. A neutron sou hanging in the centre, for observation; measuring leads (attached near the exterior. The apparatus was still a litt small to sustain a fission reaction independently, but a

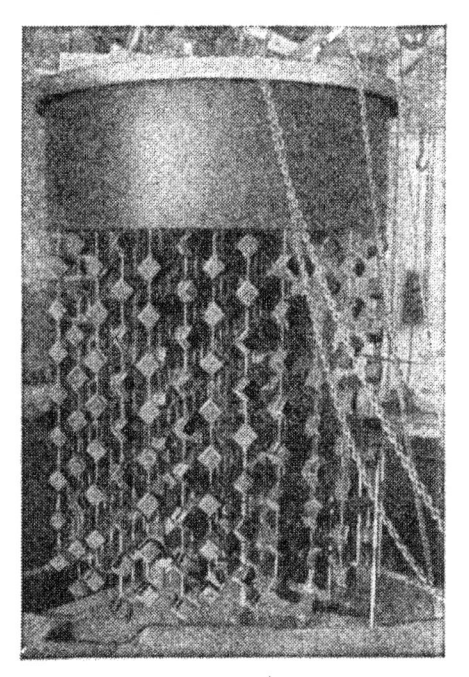

Figure 40A.—Interior of model of uranium pile.

increase in its size would have been sufficient to start o process of energy production.

The first uranium reactor large enough to yield energy built in Chicago, under the direction of Fermi, in 1942.] built of uranium and graphite and began to react in Dece 1942.

When the chain reaction has begun, the uranium pile up nuclear energy in the form of heat: the uranium simpl hot. If it is desired to utilize the energy further technically, the heat must be removed in some way. This involves a number of technical problems which have still not been solved satisfactorily enough to permit the construction of an economically perfect power station based on nuclear energy. But this is just a question of time; sooner or later there are bound to exist

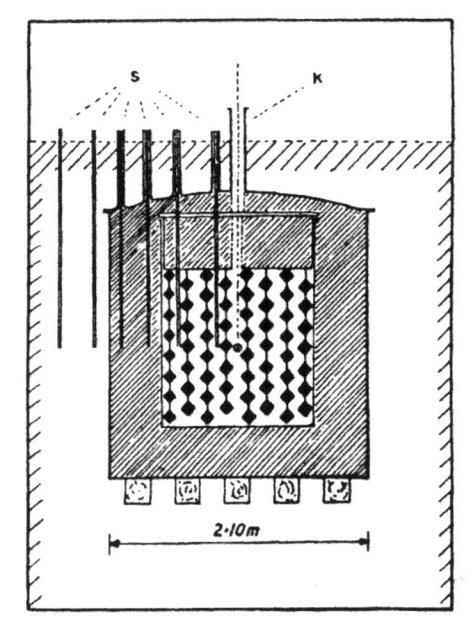

Figure 40B.—Plan of uranium pile.

power stations, stations for generating and transmitting heat, and naval engines, driven by nuclear energy.

The uranium reactor has a feature which makes it difficult to use it as a source of energy, but on the other hand, makes it very useful in another respect: It is filled by an extraordinarily intense radioactive radiation, which is extremely dangerous to every living being in the vicinity of the reactor. For this reason, the reactor is shielded by walls many metres in thickness, constructed of concrete or a like material. The reason for the presence of these radiations is obvious: The interior of the uranium reactor is the scene of nuclear reactions, occurring on a vast scale, which produce radiation of all kinds (alpha, beta and gamma radiation). An effect, in particular, of the gigantic neutron intensity in the reactor is that substances inserted into the apparatus very quickly become radioactive by the processes discussed in Section 4 of our Sixth Lecture. Thus the uranium reactor can be used as an especially efficient 'crucible' for artificially radioactivated substances. In fact, this has hitherto been the technically most important application of the uranium pile. For instance, plutonium, so important today as an 'explosive' for military purposes, is produced exclusively by a transmutation of ordinary uranium in the uranium reactor. This brings us next, after the question of the generation of energy, to the second main problem of nucleonics, which may be called 'ennoblement of matter'.

IV. ENNOBLEMENT OF MATTER BY NUCLEAR REACTIONS

More valuable materials are to be produced from those of a lesser value. New substances can be produced by nuclear reactions in very small amounts only. Therefore, their production is worth while only when they happen to be especially valuable. These extremely precious materials, which represent a considerable value, even in small quantities, are the radioactive materials. Their value consists in their radiation, which can be utilized in many ways and is considerable even when the quantity of the substance is very small. For this reason, the most important application of nuclear physics at the present time consists in an artificial production of radioactive materials.

Radioactive substances can be used for various purposes. They have been used, in medicine, for several decades, to irradiate malignant tumours, which experience has proved to be far more seriously injured by radioactive emissions than healthy tissue. In the majority of cases, of course, x-rays are used for this purpose. But whenever a complication arises that makes it difficult to approach the diseased spot without injuring other tissue, the use of radioactive preparations is preferred. However, since the natural radioactive substances are available for medical purposes in very limited quantities only—apart from radium itself, only *mesothorium*, discovered by O. Hahn, comes into consideration at all—it is hoped to achieve important progress in medical research by the production of artificial radioactive materials in sizable amounts, especially those which possess chemical properties different from those encountered in the natural ones.

Radioactive substances have still another use: Quite minute quantities of them are mixed with luminescent materials, which give out a constant luminescence due to the effect of radiation. Their best known application is in the luminous dials and hands of watches.

The radiation of radioactive preparations is used also for the testing of industrial materials for internal, structural imperfections, to which use x-rays are customarily put. The gamma radiation is the one utilized for this purpose. This procedure is applied especially to pieces too thick to be penetrated by x-rays, but which will still let gamma radiation pass through. This method of testing industrial materials has the great advantage over other procedures that it does not injure the material itself.

Let us now discuss the production and utilization of radioactive substances in a little more detail.

In order to produce artificial radioactive substances, a suitable substance is irradiated by neutrons in a uranium reactor, or by protons or deuterons in a high-voltage generator or cyclotron. A certain practical difficulty is encountered here, owing to the fact that the substance which it is desired to produce is often present in the original material in the form of minute, imponderable impurities. This substance may be chemically different from the original substance or it may be identical with it, that is to say one of its own unstable isotopes. When the chemical properties of the two substances are different, and the substance is present in ponderable quantities, the two can always be separated by chemical means without the slightest difficulty. But the situation is different when dealing with imponderable quantities, as in the cases under consideration here. In these cases, adsorption phenomena frequently occur and prevent the application of the customary chemical processes of separation. The problem can often be solved by adding in advance a considerable amount of a non-radioactive isotope of the substance to be produced, to the original substance. The adsorption phenomena will then play just a negligible part, and the radioactive substance precipitates together with the stable isotopes.

The chemical processes which can be applied when handling short-lived radioactive isotopes, have been developed and perfected to a high degree by O. Hahn and his associates, in particular.

One of the most important artificial radioactive substances is radioactive phosphorus. For instance, carbon bisulphide (CS_2) is bombarded by neutrons. The following reaction will occur in the sulphur atoms:

$$_{16}S^{32} + {_0}n^1 \longrightarrow {_{15}}P^{32} + {_1}H^1$$

The sulphur atom of mass number 32, which constitutes approximately 32 per cent. of ordinary sulphur, and the neutron produce a radioactive phosphorus atom of the same mass number, plus a proton. The half-life of this radioactive phosphorus is relatively long, 14·5 days, which is obviously an advantageous feature as regards its practical utilization; it emits an electron and changes back into the original sulphur atom—viz.:

$$_{15}P^{32} \longrightarrow {_{16}}S^{32} + {_{-1}}e^0$$

According to a suggestion by Erbacher, radioactive phosphorus is obtained simply by diluting the irradiated carbon bisulphide with water; the radioactive phosphorus is dissolved, in the form of ions, in the water which is then separated from the carbon bisulphide by one method or another.

The situation is more difficult where the radioactive substance is chemically identical with the parent substance. In such a case, it would be logical to expect a separation to be impossible. Nevertheless, it is feasible under certain circumstances; namely, when it is simply a case of the absorption of a neutron, the excitation energy is emitted in the form of a photon, as gamma radiation. But the latter produces a recoil in the nucleus which may cause the atom subsequently to become electrically charged, or to be torn out of its chemical bond. In this case, skilful chemical operations will effect the separation of the radioactive atoms from the other atoms. Such methods have been developed by *Szilard* and *Chalmers*, and others.

V. ARTIFICIAL RADIOACTIVE SUBSTANCES AS TRACERS

We have mentioned a few ways of utilizing artificial radioactive substances. But there is still another application of these substances which has been widely utilized during recent years, and which may be said to be the most important one at the present time. This application consists in the use of radioactive atoms as *tracers*. This is what is meant: Formerly, the identification, at some later instant, of individual atoms of a particular element was quite impossible. The reason for this was that the path followed by these atoms in biological or chemical processes could not be followed in detail because of the presence of other atoms of the same kind in the substance or organism being studied. Now, however, it is possible to attach a label to any element—as a ring is fastened to a leg of a carrier pigeon or migratory bird. This label is radioactivity which enables the path of the element to be followed in all its details.

We may illustrate this method by a simple example: Let us assume that we wish to study the diffusion of the atoms of a solid substance within the substance itself—for example, the diffusion of lead atoms in lead. Prior to the discovery of radioactivity, this would have been impossible, for one would never have been able to recognize an individual atom or to distinguish it from other lead atoms. But today, if a piece of lead containing radioactive atoms is brought into close contact with another piece which does not contain any, the lead atoms will be reciprocally interchanged between the two pieces by diffusion, and gradually increasing numbers of radioactive atoms will be discovered in places where originally there was no radioactivity at all. In this way we can obtain information concerning the velocity with which lead atoms diffuse in solid lead.

Let us take another example, the enormous practical value of which will perhaps be even more illuminating: In testing the filter of a gas mask, the main object is to determine to what degree it absorbs the poisonous substances, against which it is intended to serve as a protection. This can be accomplished quite simply in the following manner: Radioactive atoms of one of the elements contained in the poisonous substances are added to the latter, and the substances are then sent through the filter. These radioactive atoms undergo the same chemical reactions as the stable ones. After the poisonous substances have passed through the filter, one merely has to determine whether or not radioactivity appears at the other end of the filter, and if so, the intensity of this radioactivity will indicate what percentage of the poisonous substance has passed through the filter. The individual parts of the filter can also be checked, since, after the poisonous substance has passed through, the intensity of the radioactivity acquired by the parts in question by adsorption of the poisonous substance can be determined. This procedure will furnish information concerning the efficiency of the individual parts of the filter. Similarly, it is possible to check whether or not the rubber covering of the gas mask is actually hermetically sealed against the poison gas, by bringing the latter in contact with one side of the rubber covering and observing whether any radioactivity can be detected on the other side. If so, the rubber covering is proved to be unsuitable for a gas mask. Such testing procedures have been described by Born and Zimmer, and are actually being used.

VI. ARTIFICIAL RADIOACTIVE SUBSTANCES IN CHEMISTRY

In chemistry, artificial radioactive substances are used as tracers on an ever increasing scale. Let us give first an example from the field of quantitative analysis: Erbacher and Philipp attempted a quantitative analysis of a mixture of gold, iridium and platinum. After a reduction by hydrogen peroxide, the gold was precipitated in its metallic form, and was then weighed in order to determine whether its quantity was sufficiently close to the original quantity of gold introduced into the mixture—in other words, whether the gold has actually been separated quantitatively. This actually seemed to be the case—the separation appeared to have been completely successful. As a countercheck, a small amount of radioactive gold was added to the gold. It was found that the radioactivity of the separated quantity was perceptibly less than that of the original quantity. This proved that the gold had not been separated quantitatively, and that the seemingly perfect result had been due to the fact that a quantity of platinum and iridium had accompanied the precipitated gold, and the quantity of this platinum and iridium happened to be exactly equal to the missing quantity of gold.

This example reveals what the important factor is in the utilization of radioactive substances in quantitative analysis. A small quantity of a radioactive isotope of a substance is added, as a tracer—as a label or tag, as it were—to the substance itself. Then, the radioactivity of the tracer enables us to follow the progress of the substance throughout all its reactions, and all we need know is the half-life of that radioactive isotope to determine the quantity of it present at any given moment. The measurement of the intensity of radioactivity gives a result no less true than that which we should get by weighing the substance itself—in fact, as the above example indicates, an even more reliable result in many an instance, since it will show infallibly whether or not the substance in question is actually the one sought.

Secondly, in chemistry it may be necessary at times to investigate exchange processes, which used to be inaccessible to every method of investigation, particularly where it is a matter of an interchange of elements having the same properties between substances. The question to be answered may be, for instance, whether the sulphur atoms in sulphuric acid ions and sulphurous acid ions are interchanged between these two substances when they combine. Up to now, the difficulty consisted in the fact that it was impossible to distinguish the atoms of one of these two substances from the atoms of the other. But today, radioactivity affords the possibility of labelling at least some—but in any case a sufficient number—of the atoms of one of the two substances. If after the subsequent separation of the two substances, the atoms thus labelled are found to be present in the other substance, this will prove that atoms have been interchanged. Such experiments have shown that an interchange of sulphur atoms actually occurs between the SO''_4 and SO''_3 ions. (The two apostrophes after the symbol indicate the double negative charge of these ions.)

There are some further chemical applications I wish to mention here very briefly. The following table gives some idea of the various possibilities of application. Let us begin to consider inorganic chemistry and start with the study of new chemical elements.

There are certain elements, the places of which in the table of the periodic system were vacant till quite recently. These are elements which—to use the customary mode of expression—had not yet been discovered but were assumed to occur in nature. The best known among these are the elements of the atomic numbers (in other words, nuclear charges) 43 and 61. It was long believed that element 43 had been discovered in nature, and it was named *masurium*. But today we have every

Nuclear Transmutations and their Application in Chemistry

Substances

Natural Radioactive Elements
Formation from Uranium and Thorium Minerals

Artificial Radioactive Elements

Formation by Radiation of Natural Radioactive Elements

Formation by Radiation from Artificial Sources
High Tension Plants: Van de Graaff; Generators: Cyclotron; Uranium Pile

Means of Detection
Electroscope; Geiger-Müller Counter; Wilson Cloud Chamber; Photographic Plates

Their Applications

Inorganic Ch.	Analytical Ch.	Synthetic Ch.	Physical Ch.	Technical Ch.	Colloid Ch.	Biochemistry
Properties of New Elements; Determination of Solubility and Constitution; Lattice Structure	Efficacy of Chemical Dissociation; Quantitative Analysis; Non-destructive Analysis	Discovery and Manufacture of BiH_3	Speed of Reaction; Absolute Surfaces, Surface and Structural Changes; Processes of Diffusion or Dispersion; Reactions in Solid State	Material Testing by γ-Rays; Radioactive Photogenic Masses Test of Permeability to Gases	Detection of Colloidal & Crystalloidal Solutions; Ageing of Sols and Gels	Metabolic changes in Plants, Animals and Human Beings; Formation of Bones and Teeth; Therapeutic Tests

reason to believe that this was an error, and that no stable element of this kind can exist. For in the meantime every isotope of this element of appreciable stability has been produced artificially, and every one of them has been proved to be radioactive. The mass number of the longest-lived isotope is 99; its half-life is approximately four million years. Since this is a short period of time relative to the age of the earth, element 43 cannot possibly occur in nature in any measurable quantity. Since this element can be produced artificially in the uranium reactor, it has recently been renamed *technetium* (symbol: *Tc*). The above mentioned most stable isotope is therefore designated by $_{43}Tc^{99}$. The state of affairs is similar for element 61. This element, too, was believed to have been discovered in certain minerals, and it was named *illinium*. But, again, the discovery could not be confirmed. It is practically certain that neither of these elements exists in stable form.

But since it is possible to produce these elements, in radioactive varieties at least, it is possible also to produce chemical reactions with them. Radioactivity does not interfere in the least with chemical reactions. Thus, the chemical properties of element 43 have been investigated in a whole series of experiments. The investigation of element 61 is more difficult, because it is one of the rare earth elements, and its chemical properties are almost identical with those of the other members of that group of elements.

The other two missing elements, those of atomic numbers 85 and 87, have also been produced artificially. Element 85 was produced by Corson, McKenzie and Segré from bismuth, by bombardment with alpha rays of high energy (32 Mev.), and was named *astatine* (symbol: *At*). It is formed by the following reaction:

$$_{83}Bi^{209} + {}_2He^4 = {}_{85}At^{211} + {}_0n^1 + {}_0n^1$$

Since that time, traces of $_{85}At^{218}$ have been detected, by Karlik and Bernert, among the natural radioactive substances, as a product of the decay of $_{84}RaA^{218}$ ($_{84}Po^{218}$).

Element 87 is a product of the decay of a neptunium isotope. Traces of it have also been detected by Perrey in the radioactive decay of natural actinium. It has been named *francium*.

Finally, the periodic system of elements, which used to end with uranium, has been extended artificially to atomic number 96. We have already mentioned neptunium and plutonium. These two elements are produced principally in the uranium reactor, in the way indicated in the following formulae:

$$_{92}U^{238} + {}_0n^1 = {}_{92}U^{239} \longrightarrow {}_{93}Np^{239} + {}_{-1}e^0$$
$$_{93}Np^{239} \longrightarrow {}_{94}Pu^{239} + {}_{-1}e^0$$

But other isotopes of these elements have also been produced artificially.

Elements 95 (*americium—Am*) and 96 (*curium—Cm*) have been obtained as follows:

$$_{92}U^{238} + {}_2He^4 = {}_{94}Pu^{241} + {}_0n^1;$$
$$_{94}Pu^{241} \longrightarrow {}_{95}Am^{241} + {}_{-1}e^0; \quad {}_{94}Pu^{239} + {}_2He^4 = {}_{96}Cm^{242} + {}_0n^1$$

The production of new chemical elements is therefore no longer a dream of the future, but an important part of modern nucleonics.

Now we come to synthetic chemistry, the science of the building of new chemical compounds. Bismuth hydrate will be a good example. It had been concluded by chemical analogies that the production of such a compound was bound to be a distinct possibility. But owing to the extreme difficulty of detecting this gas, all attempts to produce it seem to have failed. However, this project has been crowned by success, thanks to the use of radioactive bismuth as a tracer.

It may be mentioned too that in colloid chemistry room for further applications has been found in connection with the detection of colloidal and crystalloid solutions, as well with the ageing of sols and gels.

VII. ARTIFICIAL RADIOACTIVE SUBSTANCES IN BIOLOGY AND BIOCHEMISTRY

Artificially radioactivated substances have also been used, with great advantage, as tracers in biology. In a living organism, changes are often far slower than in a test-tube. This fact calls primarily for the use of radioactive substances with longer half-lives.

One of the most important applications in this field has been the study of metabolism, carried out by Hevesy and others. Formerly, it was only possible to determine the overall picture of a metabolism by verifying what quantities of a certain substance, introduced into the organism for this specific purpose, were still present in its various individual parts after a certain length of time. However, it was impossible to distinguish this particular substance from the chemically identical one previously present in the organism. This circumstance resulted in considerable uncertainty, particularly as regards the speed of the distribution of the substances introduced among the various organs. But the technique of labelling atoms by their radioactivity quickly eliminated this difficulty. This tracer technique enables us to distinguish the atoms which have been deliberately introduced into the organism from those which were there previously.

For instance, Born, Schramm and Zimmer grew tobacco plants in a nutrient soil containing a substance enriched with radioactive phosphorus, which was absorbed by the plants, since phosphorus is one of substances essential for the maintenance of organic life. The progress and paths of the phosphorus in the plants could be observed, and the points of its strongest concentration could be recognized. The major part of the radioactive phosphorus passed into the uppermost, youngest still growing leaves, while less found its way into the lower leaves, and still less into the fully developed leaves. One leaf, in which the circulation of sap had altogether ceased, absorbed no phosphorus at all. Another study of a similar nature determined the speed of travel of the phosphorus in the plant; it was found to be about 10 cm. per second.

Artificially produced radioactive substances, and, again, radioactive phosphorus in particular, have been used to study animal metabolism. In these cases, the phosphorus was mixed with the food of the animals or injected into their bodies (*Hevesy*). This technique makes it possible to determine, after a while, in what parts of the organism the phosphorus has a tendency to settle, and also the speed of its elimination, so that information concerning metabolism can be obtained not only numerically and as to proportion, but also as to details. Among other things, it has been ascertained that after a certain length of time, the phosphorus settles principally in the bones and in the liver, and after another period of time, in the teeth. A great many important biological data can be obtained in this manner.

The quantities of radioactive substance required for such experiments are so minute that they cannot result in any harm to the organism.

Another important problem which could be tackled by the tracer method was that of the assimilation by plants of carbon dioxide. It is a well-known fact that green plants utilize the effect of sunlight (in other words, a photochemical process) to assimilate carbon dioxide from the air and convert it into hydrocarbons. This is how they store up solar energy. But very little was known about the details of this process, and various theories were formulated about its mechanism. It was known for a certainty that every plant must absorb approximately four photons of light so as eventually, with the aid of their energy, to accomplish the chemical reaction. In order to explain this problem, the American scientists Ruben, Hassid and Kamen used carbon dioxide containing radioactive carbon of mass number 11, which has a half-life of 20 minutes. They found that in order to prepare the way for an assimilation, first a reaction in darkness takes place, in the course of which both the carbon and the oxygen of the carbon dioxide (CO_2) are combined with the addition of hydrogen in the form of the carboxyl group (of the residue COOH) in a giant organic molecule. Later on, glucose (grape sugar, $C_6H_{12}O_6$) is formed from the carboxyl group, possibly by the roundabout way of larger sugar molecules. As a result, the important fact was established that the process of assimilation takes place in several stages. However, we shall not enlarge on further details of this subject here.

VIII. ARTIFICIAL RADIOACTIVE SUBSTANCES IN MEDICINE

Artificial radioactive substances are also destined to render valuable service at a later date in medicine. In contrast to the experiments described above, where the important thing is the investigation of the normal unimpaired organism, by the aid of small quantities of radioactive materials, it is also possible to test the effect of larger quantities on the organism by virtue of their radiation. A comprehensive investigation was carried out along these lines by Scott and Cook, who introduced radioactive phosphorus, produced in the Berkeley cyclotron, into the food of young hens, and then studied the haematological changes produced by the radiation. They discovered a number of interesting changes and compared them with others produced by x-rays. It is well known that blood contains two kinds of corpuscles—red and white ones. The latter are again divided into several groups, the most important of which are the polymorphonuclear leucocytes, the eosinophilic leucocytes and the basophilic leucocytes. They differ from each other in size, internal structure and staining by various substances. x-rays first make the number of lymphocytes decrease and the number of polymorphonuclear leucocytes increase, but after a short while, this multiplication of the leucocytes falls off. On the other hand, the radiation of the radioactive phosphorus introduced into the body does not affect the lymphocytes to any significant extent, but produces, instead, an appreciable permanent decrease in the number of the polymorphonuclear leucocytes. Furthermore, the corpuscles known as monocytes and the eosinophilic leucocytes are affected a little, but not strongly. The eosinophilic leucocytes and the red blood corpuscles show a slight increase in number.

This specific effect encouraged the American scientists to try to treat certain forms of leucaemia—in which the diagnosis of the disease showed considerable haematological changes—by doses of radioactive phosphorus. The phosphorus settles predominantly in the bones, and as it is well known, the red corpuscles are formed in the bone marrow, so that radioactive phosphorus is evidently capable of affecting their formation. We see how different is the effect of the radioactive phosphorus from that of x-rays, which penetrate all tissues to the same extent. The first experiments in this field are said to have been promising. In consequence of the war, however, no further information could be obtained about them. Similar experiments designed to study haematological changes have been carried out in Germany too.

Furthermore, experiments have been made with injections of radioactive lead, partly in view of the cases of lead poisoning occasionally occurring among workers, and partly with the hope of achieving therapeutic results. It was discovered that when lead is introduced into an organism, most of it is excreted very rapidly, and only a small residue remains in the liver and in the kidneys. Lead does not settle in cancerous tissue either, so that all attempts to use it for the treatment of cancer are doomed to failure. On the other hand, bismuth settles very rapidly in diseased tissue. An attempt to treat cancerous tissue by radioactive bismuth may possibly prove successful.

But all these attempts are still in their initial stages, and many years are bound to elapse before the experimental investigation emerges from the domain of pure abstract research into that of practical therapy. It would of course be foolhardy to attempt to apply these methods to the human body before they have been tested sufficiently.

Another possible application of artificially produced radioactive substances is in the investigation of functional disturbances of organs. Many of the organs, both of human and animal bodies, have not just one but sometimes several functions. If such an organ is injured, it is often difficult to ascertain which of its functions is impaired and which is still carried out normally. Since every function is associated with a different type of metabolism, it is possible, for instance, to introduce into the organism certain radioactive substances, which are specific for certain functions, and to observe whether or not the organism does with them what it is supposed to do. It is possible by means of this technique to distinguish the impaired functions from the unimpaired ones. A new method of medical diagnosis may develop along these lines.

IX. THE USE OF STABLE ISOTOPES

Among the rare isotopes of the various elements, a special part is played by deuterium, the hydrogen isotope of mass number 2, since the ratio of its mass to the mass of the common hydrogen isotope of mass number 1 is far greater than usually found among isotopes of other elements. For this reason, there are noticeable differences in the chemical properties of these two isotopes, which make it easy for us to detect them when they are together. Consequently heavy hydrogen can also be used as a tracer. For instance, fatty acids have been built up with heavy hydrogen, instead of the common hydrogen isotope, and these fatty acids have been introduced into an organism, and the problem studied as to how the organism makes use of them. This experiment revealed that the long-chained fatty acids settle in the liver and in the fatty tissues, whereas the short-chained ones are consumed immediately. The experiment would not have been possible with ordinary hydrogen, because such fatty acids are always present in the organism, and therefore it would have been impossible to distinguish the fatty acids deliberately inserted into the food from those previously present in the organism. Similar experiments have been performed with nitrogen of mass number 15 and oxygen of mass number 18.

In conclusion, let us mention one more application of a nuclear reaction to a branch of physics itself—to optics. With the cyclotron, the opposite of the old dream of the alchemists—not a conversion of mercury into gold, but one of gold into mercury—has been achieved. Gold is monoisotopic, in other words, it has only one stable isotope, $_{79}Au^{197}$. If a nuclear reaction is produced in gold, as expressed by the formula:

$$_{79}Au^{197} + _{0}n^{1} \longrightarrow {_{79}Au^{198}} \longrightarrow {_{80}Hg^{198}} + _{-1}e^{0}$$

which indicates the emission of an electron, it will produce only one of seven stable mercury isotopes, six of which occur in ordinary mercury in practically equal proportions. This mercury isotope is very suitable for certain optical researches. Namely, when ordinary mercury vapour—in other words, the natural mixture of mercury isotopes—is caused to glow by an electric charge, the spectra of the individual isotopes vary but very slightly, and their spectral lines become superimposed, in what we call a *fine structure*. This fine structure is absent in the mercury produced from gold, and therefore this particular mercury isotope is most suitable for standard spectroscopic measurements (as was proposed some years ago by W. E. Williams), where the essential aim is to obtain lines as sharply defined as possible. If we recall the purpose mentioned at the beginning of this chapter, the production of more valuable substances out of those of lesser value, we shall appreciate that it is an interesting indication of the transmutability of all things that in this particular instance mercury is more valuable than gold, instead of vice versa.

This concludes our survey of the applications of nuclear physics. All that we have discussed represents merely the beginnings of a development, the future progress of which cannot even be estimated. But practical applications are not the most important aspect of nuclear physics, and that is why they have not been discussed in greater detail in these lectures; the practical benefits of a knowledge of natural phenomena constitute a later problem. For the time being, the main thing is to understand the structure of the nucleus of the atom. The purpose of these lectures has been to present a summary of what has been accomplished in this field and what still remains to be done. All that has been described here was intended to impart both to listeners and readers something of the magical effect on all of us of those natural phenomena which are so difficult of access, and in particular of those whose internal laws we have not yet fathomed.

APPENDIX

RESEARCH IN GERMANY ON THE TECHNICAL APPLICATION OF ATOMIC ENERGY*

Even ten years ago, physicists were well aware that the utilization of atomic energy could not be realized without a fundamental extension of scientific knowledge. In spite of the remarkable progress in experimental nuclear physics which followed the introduction of high-voltage equipment and the invention of the cyclotron, no physical phenomenon was known, even as late as 1937, which offered the remotest possibility of exploiting the enormous quantities of energy lying latent in atomic nuclei.

It was the discovery of the fission of uranium by Hahn and Strassmann[1] in December 1938—in other words, the fact that the uranium nucleus can be split into two fragments of comparable mass when bombarded by neutrons—which brought the actual utilization of atomic energy within reach. Following this discovery, Joliot and his co-workers[2] succeeded in proving, in the spring of 1939, that in the act of fission the uranium nucleus itself emits several neutrons, thus making a chain reaction fundamentally possible. Thereafter the possibility of nuclear chain reactions was eagerly debated among physicists, particularly in the United States; in Germany it was discussed by Flügge[3] in *Die Naturwissenschaften* in the summer of 1939. Meitner and Frisch[4] had already directed attention to the enormous quantities of energy set free in the fission process.

Public interest in the problems of atomic physics was

* This article is a slightly abridged translation of a paper which appeared in *Die Naturwissenschaften* and also in *Nature*, vol. 160, page 211, Aug. 16, 1947.

negligibly small in Germany between the years 1933 and 1939, in comparison with that shown in other countries, notably the United States, Britain and France. Thus, while in America, previous to 1939, a whole series of modern research laboratories equipped with high-voltage plants and cyclotrons was springing up, in Germany there were only two adequately equipped laboratories; and these were not supported by the State, but sponsored by a private body, the Kaiser Wilhelm Gesellschaft. These two institutes were the Kaiser Wilhelm Institutes at Heidelberg and Berlin-Dahlem; each possessed a small highvoltage set suitable for nuclear research. A cyclotron for such work was altogether lacking—the Heidelberg cyclotron, again built entirely by private funds, and mainly designed for medical investigations, was started as late as 1938, and could not be tested out before 1944. Only with the outbreak of war did the awakened interest of the authorities allow of more extended facilities for nuclear research.

The following report deals wih those particular investigations which had for their purpose the technical utilization of atomic energy. The many purely scientific problems which arose in more or less close connection with the technical problem will not be discussed here; they will ·be dealt with in a forthcoming FIAT* Review by Flügge and Bothe. I may, however, mention the extensive chemical investigations of Hahn and his coworkers on the fission products of uranium, carried out throughout the War in the Kaiser Wilhelm Institute for Chemistry, the great majority of which have been published.

Almost simultaneously with the outbreak of war, news reached Germany that funds were being allocated by the

* FIAT Reviews of German Science, 1939–1946: a series of authoritative accounts of the progress made in both natural and applied science in Germany during the War. These reviews have been sponsored jointly by British, American and French FIAT (Field Intelligence Agency (Technical)), and a limited edition is expected to be ready for distribution before the end of 1947.

American military authorities for research on atomic energy*. In view of the possibility that England and the United States might undertake the development of atomic weapons, the Heereswaffenamt created a special research group, under Schumann, whose task it was to examine the possibilities of the technical exploitation of atomic energy. As early as September 1939 a number of nuclear physicists and experts in related fields were assigned to this problem, under the administrative responsibility of Diebner. I should mention the names of Bothe, Clusius, Döpel, Geiger, Hahn, Harteck, Joos and v. Weizsäcker among those so employed. At Schumann's behest, the Kaiser Wilhelm Institut für Physik in Berlin-Dahlem was nominated as the scientific centre of the new research project. The Institute came accordingly under the administration of the Heereswaffenamt; a step which disregarded the rights of the Kaiser Wilhelm Gesellschaft, and so led to the departure of its director, Debye, who as a Dutch citizen could not continue to serve under the aegis of a German war department.

As a result of the first conferences in the autumn of 1939, it was clear that there were two lines of attack possible in the exploitation of nuclear energy. One could attempt the separation of the rare isotope U(235) from ordinary uranium. This isotope, following theoretical arguments due to Bohr, must be immediately applicable either to the controlled production of energy, using primarily the slow-neutron reaction, or directly as an explosive in bombs, using the fast-neutron reaction; the separation of U(235) was, however, a problem which made the greatest possible demands on engineering technique. Secondly, one could mix ordinary uranium with a substance which would slow down the neutrons produced in the nuclear fisson without absorbing them. These slow neutrons give rise preferentially to

* Cf. Smyth Report, p. 27: Initial approaches to the Government. In actual fact, U.S. Government funds were first used at the turn of the year 1939–40 (p. 28), whereas the first discussions between men of science and the American navy took place as early as March 1939.

the fission of U(235) and thus maintain the chain reaction. A rapid deceleration of the neutrons is required, in order that the region of resonance absorption by U(238) should be rapidly traversed; for if absorbed they are lost to the chain reaction. The advantage of this arrangement is that the chain reaction can be controlled through the heat developed thereby, so that energy can be abstracted in amounts sufficient for technical applications.

Thus two lines of purely scientific investigation were marked out: first, to develop refined methods for the separation of isotopes; secondly, by measurement of the effective cross-sections of a range of possible substances, to determine whether the alternative line of attack was at all practicable. Harteck pointed out, as early as the autumn of 1939, that it might be advantageous, in regard to the second scheme, to have the so-called moderator physically segregated from the uranium. This suggestion gave rise to theoretical investigations as to whether, with the effective cross-sections of such moderator substances as were known at the time, an arrangement having a homogeneous mixture of uranium and moderator, or one with a local separation (for example, in layers), led to the more favourable production of energy. A tentative theoretical investigation made by Heisenberg, in December 1939, led to the result that while ordinary water was unsuitable as a moderator, it should be possible with heavy water (D_2O) or very pure carbon to produce energy in positive amount, provided the moderator and uranium were arranged in layers. This arrangement, however, demanded the highest degree of purity of the substances involved. At the same time, it was evident that a certain minimum size of apparatus was necessary for the production of energy. Nevertheless, with a small set it is still possible to determine whether there would be a production of energy if the apparatus were suitably enlarged. Thus if we feed such a small plant with neutrons from an internal source, more neutrons must escape from the surface than are supplied by the source, if the layer arrangement is favourable to energy production; if unfavourable, then fewer neutrons escape than are supplied by the source. These small model plants, which are continuously fed from a neutron source, are called 'neutron-injected piles'. The ratio k of the number of neutrons escaping from the pile to that fed in by the source can be used to characterize the pile. If $k < 1$, the arrangement is unsuitable for the production of energy; if $k > 1$, energy will be produced on enlarging the pile.

In 1940, measurements of the most important effective cross-sections were carried out, especially by Bothe and his collaborators, and by Döpel and Heisenberg. At the same time, investigations on the masses and energies of the fission products were being pursued by Jentschke and Prankl[5] and by Flammersfeld, P. Jensen and Gentner[6], and on the spectrum of the neutrons produced by Kirchner and v. Droste and by Bothe and Gentner[7]. The theory of neutron absorption in the U(238) resonance line was laid down by Flügge and Heisenberg. On the technical side the following results were the most important: the absorption cross-section of heavy water proved to be so low that this substance was certainly usable in the construction of a uranium pile (Döpel and Heisenberg). The work of v. Droste on large quantities of sodium uranate, and of Harteck and the Hamburg group—Groth, H. Jensen, Knauer and Süss—on U_3O_8 in solid carbon dioxide, furnished the first criteria for the distribution of neutron density in certain arrangements of uranium and moderator. In the autumn of 1940, the first pile, constructed of layers of U_3O_8 and light paraffin, was built at Berlin-Dahlem and its characteristics measured (Wirtz, Fischer, Bopp). This model pile gave, as expected, $k < 1$, that is, the arrangement was not suitable for the production of energy. Nevertheless, it yielded valuable data for further piles using alternate layers of U_3O_8 and heavy water.

In the summer of 1940, v. Weizsäcker pointed out that a uranium pile, besides generating the fission products of uranium, will constantly produce the uranium isotope U(239) and its transformation series; and that theoretically these transformation products should show the same properties as U(235) in regard to neutron fission. It remained thereby an open question whether β-decay ended at element No. 93 or 94 or even later; for, since no cyclotron was available in Germany, these elements could not be prepared in sufficient quantity for the examination either of their nuclear properties or chemical characteristics. Nevertheless, it appeared likely, from v. Weizsäcker's work, that an energy-producing pile might be used for the production of an atomic explosive, even though the practical details involved remained uncertain. In fact, this method had been employed on the grand scale in America. The American piles deliver as a transformation product of U(239) the element plutonium $_{94}Pu^{239}$, which is used in the manufacture of atomic bombs.

Important technical problems immediately arose out of these scientific investigations. The production of U_3O_8 of the highest purity was assigned to the Auer-Gesellschaft by the Heereswaffenamt. The casting of the corresponding metal powder was afterwards allotted to Degussa in Frankfurt. The production of heavy water, which was obviously of the greatest importance for the construction of a uranium pile, was planned at the Norsk Hydro factory at Rjukan in Norway. Harteck, in conjunction with Süss, H. Jensen and Wirtz, developed a number of projects which resulted in an increase of heavy water production at Rjukan far beyond the former output of 10–20 litres a month. Moreover, Harteck and Clusius put forward detailed plans for the production of heavy water in Germany. The improvements in the Norsk Hydro factory finally increased production in the summer of 1942 to about 200 litres per month. Further, steps were taken by order of the Heereswaffenamt for the production of very pure carbon. The attempts to exceed the degree of purity afforded by the best technical electrographite failed, however, for the time being.

The most important progress in the uranium project was achieved during the year 1941. Initially some negative results were recorded. Thus, the enrichment of U(235) by the Clausius–Dickel thermal diffusion method, using gaseous UF_6, proved impossible (Fleischmann, Harteck and Groth). The absorption cross-section for neutrons of the highest purity electrographite was determined in the Kaiser Wilhelm Institute at Heidelberg (Bothe and Jensen[4]), and the behaviour of pure carbon estimated from the results. It appeared, according to the information then available, that even the purest possible carbon was unsuitable for the construction of a uranium pile; whereas, as is well known, carbon has been used in the United States with complete success. Whether the Heidelberg conclusions were falsified by insufficient consideration of the chemical impurities present in commercial graphite (for example, hydrogen or nitrogen), or by deficiencies in the theory, can scarcely be assessed for the moment. In any event, the Heidelberg experiments on graphite and beryllium (Fünfer and Bothe[5]) made it clear, in connection with later experiments in the Berlin–Dahlem Institute, that both pure carbon and pure beryllium were highly suitable for use as an outer cover for a uranium pile, since their low absorption cross-section and high reflecting power restrict the spread of the neutrons escaping from the pile, thus reducing its minimum dimensions.

In the summer of 1941, 150 litres of heavy water were available for the first experiments on a neutron-injected pile built up of uranium and heavy water (Döpel and Heisenberg, Leipzig). The uranium and heavy water were arranged in alternate spherical layers with the neutron source at the centre. The oxide U_3O_8 which was first employed produced only a slight increase in the number of escaping neutrons, which could scarcely be considered a clear proof that $k > 1$. The use of pure uranium metal, however, gave such a decided improvement that no further doubt of a real increase in the number of neutrons ($k > 1$) was possible (about February or March 1942).

Here then was the proof that the technical utilization of atomic energy was possible, and that the mere enlargement of the Leipzig apparatus must furnish an energy-producing uranium pile.

At the same time, important administrative changes were taking place. At a meeting held in the building of the Reichsforschungsrat in Berlin, on February 26, 1942, the results to date were reported to the Minister of Education, Rust, and several directors of war research. The uranium project was transferred from the Heereswaffenamt to the control of the Reichsforschungsrat; and the then president of the Physikalisch-Technische Reichsanstalt, Esau, was made responsible for the project. On June 6, 1942, there was a second meeting at Harnack House in Berlin, when the results of the uranium project were reported to Speer, as Minister for War Production, and to the armament staff.

The facts reported were as follows: definite proof had been obtained that the technical utilization of atomic energy in a uranium pile was possible. Moreover, it was to be expected on theoretical grounds that an explosive for atomic bombs could be produced in such a pile. Investigation of the technical sides of the atomic bomb problem—for example, of the so-called critical size—was, however, not undertaken. More weight was given to the fact that the energy developed in a uranium pile could be used as a prime mover, since this aim appeared to be capable of achievement more easily and with less outlay. As to the separation of the uranium isotopes, no method was known which would have allowed of the production of an atomic explosive without an enormous and therefore impossible technical equipment. Incidentally, the use of protoactinium as an atomic explosive was also considered, since its nucleus is fissionable by neutrons with energies down to 10^5 eV., with the consequent possibility of a fast chain reaction. It was, however, considered to be impracticable to prepare the necessary quantities of the element.

Following this meeting, which was decisive for the future of the project, Speer ruled that the work was to go forward as before on a comparatively small scale. Thus the only goal attainable was the development of a uranium pile producing energy as a prime mover—in fact, future work was directed entirely towards this one aim. Again during the summer of 1942, discussions were held with heat experts on the technical problems of heat transfer from the uranium to the working material (that is, water or steam). Technical experts from the Navy attended the meeting with a view to the possible use of a uranium power unit in warships. The Kaiser Wilhelm Institut für Physik was restored to the Kaiser Wilhelm Gesellschaft, with the author as director. In preparation for investigations on the larger uranium piles planned in the Institute, a spacious underground laboratory was added (Wirtz).

About this time, however, the strain of the war on the already overloaded German industry was making itself felt. Uranium and uranium slugs were produced in such small quantities that deliveries were late and the larger-scale experiments were repeatedly postponed. Nevertheless, important progress was made. As early as 1941, a research group at the Heereswaffenamt (Diebner, Pose, Czulius) had made measurements on a large pile built up as a lattice of uranium cubes in a paraffin matrix; the subsequent theoretical investigation (Höcker) demonstrated that the lattice construction could in certain circumstances show advantages over the layer arrangement. An experiment made by this group with a model pile of uranium cubes in D_2O ice did, in fact, yield a larger increase in the number of neutrons than the Leipzig pile. In a later experiment, using 500 litres of heavy water, a further increase in the number of neutrons was recorded. Measurements made in the Heidelberg Institute with a small model pile defined the relation between the increase in the number of neutrons on one hand, and the thickness of the layers on the other; while experiments undertaken by Bothe and Flammersfeld at Heidelberg, and by

Stetter and Lintner in Vienna, threw new light on the fission processes occurring in U(238). A theoretical investigation by Bothe stressed the importance of the 'stopping distance' (*Bremslänge*) for the minimum size of a self-sustaining pile.

In preparation for further experiments with larger quantities of heavy water and uranium metal, the Kaiser Wilhelm Institut für Physik in Berlin began a study of the effect of graphite and water as an outer cover for the pile. Resonance absorption in uranium had been studied by Volz and Haxel and by Sauerwein. Further, the absorption cross-sections of a series of different substances were measured by Ramm, as well as by Volz and Haxel at the Berlin Technical High School. As regards the question of thermal stabilization of the energy production resulting from the temperature broadening of the resonance lines, experiments by Sauerwein and Ramm with the Berlin-Dahlem high-voltage plant were significant.

In the spring of 1943, the Norsk Hydro electrolytic plant was put out of action in a Commando raid. Its reconstruction was begun, but finally the responsible army command in Norway reported that effective protection of the plant, particularly against air raids, was impossible. In October 1943 the plant was completely destroyed in a heavy air raid. Nevertheless, about two tons of heavy water were available in Germany at the time: a quantity which, according to our calculations, was just enough for the construction of an energy-producing pile. The Reichsforschungsrat had made no effective provision for the construction of a new heavy water factory in Germany, and the pilot plant at I.G. Leuna made slow progress. It was proving, in fact, barely possible, in view of air raids and the overall strain on German production, to undertake such big building projects. The production of uranium slugs came to a temporary standstill after the raids on Frankfurt in the spring of 1944.

Even then, some progress was achieved by the Harteck–Groth–Beyerle group. As early as 1942, this group had succeeded in developing an ultracentrifuge for the enrichment of the isotope U(235). It was planned to use uranium enriched with the rare isotope in the construction of improved uranium piles, possibly in conjunction with ordinary water. At about this time, the direction of the uranium project was transferred from Esau to Gerlach. Gerlach had taken over the physics section of the Reichsforschungsrat, and he strove to promote more particularly the scientific side of the uranium problem; and at that, not only the physical, but also the medical aspect. In connection with the medical applications, the construction of a low-temperature pile, in liquid carbon dioxide, was undertaken on Harteck's suggestion. Such a pile, even of small dimensions, could be expected to yield profitable amounts of radioactive elements for tracer research, in view of the decreased absorption in the resonance lines at low temperatures.

In the winter of 1943–44, a model pile of 1·5 tons of heavy water and about the same weight of uranium plates was constructed in the Dahlem air-raid shelter through the co-operative efforts of the Kaiser Wilhelm Institutes for Physics in Berlin and Heidelberg (Wirtz, Fischer, Bopp, P. Jensen, Ritter). The number of neutrons injected from the internal source was multiplied by the factor 3, a performance approaching considerably nearer to what we called the *Labilitätspunkt*, at which the ratio k increases beyond limit and at which the uranium pile begins to radiate independently of the neutron source and thus to produce energy. The relation between neutron increase and layer thickness fulfilled expectations. Further, the stopping distances of the fission neutrons in carbon and heavy water were redetermined (Wirtz) and the former inexact measurements considerably improved. These experiments were made in the air-raid shelter of the Institute during the heaviest air raids on Berlin, and were naturally to some extent hindered by the raids. On February 15, 1944, the Kaiser Wilhelm Institut für Chemie received a direct hit. In the meantime, the Kaiser Wilhelm Institut für Physik had been partly evacuated to

Hechingen. On the instructions of Gerlach, a cellar cut out of the solid rock, situated in the village of Haigerloch, was equipped for the rebuilding of the uranium pile. It was not until February 1945, however, that the greater part of the necessary material (about 1·5 tons of heavy water, 1·5 tons of uranium, 10 tons of graphite, cadmium for the regulating rods, etc.) was finally assembled at Haigerloch, and a new pile, this time built up of uranium cubes in heavy water, with an outer cover of graphite, constructed (Wirtz, Fischer, Bopp, Jensen, Ritter). A branch of the Reichsforschungsrat at Stadtilm was allotted the remaining quantity of heavy water and a great part of the available uranium. The Haigerloch pile yielded a sevenfold neutron increase. The material available at Haigerloch, however, was just insufficient to attain $k = \infty$. A relatively small amount of uranium would in all probability have sufficed; but it was no longer possible to obtain it, since transport from Berlin or Stadtilm could no longer reach Hechingen. On April 22, Haigerloch was occupied, and the material confiscated by the Americans.

When we compare the German work reported here with the corresponding Anglo-American effort, so far as it has been made known, then the beginning of 1942 seems to be the turning point. Up to that time, both sides were dealing predominantly with the scientific problem as to whether nuclear energy could be utilized at all, and what fundamental methods had to be employed to that end. Both sides had arrived almost simultaneously at very similar results, if one excludes the field of isotope separation, in which the Anglo-Americans had made much greater progress. Furthermore, in the United States far more attention had been given to laying the groundwork for subsequent full-scale development of the uranium project; so that the first self-supporting pile was functioning as early as December 1942.

It remained to determine the technical sequel to these results. In the United States, the final decision was taken to go for the production of atomic bombs, with an outlay that must have amounted to a considerable fraction of the total American war expenditure; in Germany an attempt was made to solve the problem of the prime mover driven by nuclear energy, with an outlay of perhaps a thousandth part of the American. We have often been asked, not only by Germans but also by Britons and Americans, why Germany made no attempt to produce atomic bombs. The simplest answer one can give to this question is this: because the project could not have succeeded under German war conditions. It could not have succeeded on technical grounds alone: for even in America, with its much greater resources in scientific men, technicians and industrial potential, and with an economy undisturbed by enemy action, the bomb was not ready until after the conclusion of the war with Germany. In particular, a German atomic bomb project could not have succeeded because of the military situation. In 1942, German industry was already stretched to the limit, the German Army had suffered serious reverses in Russia in the winter of 1941–42, and enemy air superiority was beginning to make itself felt. The immediate production of armaments could be robbed neither of personnel nor of raw materials, nor could the enormous plants required have been effectively protected against air attack. Finally—and this is a most important fact—the undertaking could not even be initiated against the psychological background of the men responsible for German war policy. These men expected an early decision of the War, even in 1942, and any major project which did not promise quick returns was specifically forbidden. To obtain the necessary support, the experts would have been obliged to promise early results, knowing that these promises could not be kept. Faced with this situation, the experts did not attempt to advocate with the supreme command a great industrial effort for the production of atomic bombs.

From the very beginning, German physicists had consciously striven to keep control of the project, and had used their influence as experts to direct the work into the channels which have been mapped in the foregoing report. In the upshot they were spared the decision as to whether or not they should aim at producing atomic bombs. The circumstances shaping policy in the critical year of 1942 guided their work automatically towards the problem of the utilization of nuclear energy in prime movers. To a German physicist, this task seemed important enough. The mere possibility of solving the problem had been rendered possible by the discovery of the German scientific workers Hahn and Strassmann; and so we could feel satisfied with the hope that the important technical developments, with a peace-time application, which must eventually grow out of their discovery, would likewise find their beginning in Germany, and in due course bear fruit there.

[1] *Naturwiss.*, **27**, 11 (1939).
[2] *Nature*, **143**, 470 (1939).
[3] *Naturwiss.*, **27**, 402 (1939).
[4] *Nature*, **143**, 239 (1939).
[5] *Z. Phys.* **119**, 696 (1942).
[6] *Z. Phys.*, **120**, 450 (1943).
[7] *Z. Phys.*, **119**, 568 (1942).
[8] *Z. Phys.*, **122**, 749 (1944).
[9] *Z. Phys.*, **122**, 769 (1944).

Publication of results for which no source is cited was prohibited during the War.

TABLES

Table Ia

Physical Constants

Ionic Charge (Faraday) = 96,520 coulombs per gramme atom of univalent element.
Velocity of Light (in vacuo), $c = 2\cdot99778 \times 10^{10}$ cm. sec.$^{-1}$.
Electronic Charge, $e = 4\cdot803 \times 10^{-10}$ e.s.u. $= 1\cdot602 \times 10^{-20}$ e.m.u. $= 1\cdot602 \times 10^{-19}$ coulombs.
Rest Mass of Electron $m_0 = 9\cdot107 \times 10^{-28}$ grammes.
Ratio of Charge to Mass of Electron, $e/m_0 = 1\cdot759 \times 10^8$ coulombs per gramme.
Loschmidt's (Avogadro's) Number (Number of Molecules per Mole), $L = 6\cdot024 \times 10^{23}$.
Planck's Quantum of Action (Planck's Constant), $h = 6\cdot624 \times 10^{-27}$ ergs $\times$ sec. and $\hbar = h/2\pi = 1\cdot0543 \times 10^{-27}$ erg $\times$ sec.
Rydberg's Constant ($R = 2\pi^2 me^4/c\,h^3$) = 109,737 cm.$^{-1}$.
Mass of Electron = $5\cdot486 \times 10^{-4}$ a.m.u. (Atomic Mass Units).
Mass of Proton = $1\cdot00758$ a.m.u.
Mass of Hydrogen Atom = $1\cdot00813$ a.m.u.
Mass of Neutron = $1\cdot00895$ a.m.u.
Mass Ratio Hydrogen Atom to Electron, $M/m_0 = 1837\cdot5$.

Table Ib

Physical Units

1 Million Electron Volts: 1 Mev = $1\cdot6 \times 10^{-6}$ erg = $3\cdot83 \times 10^{-14}$ cal.
Energy Equivalent of Mass: 1 a.m.u. = $1\cdot49 \times 10^{-3}$ erg.
Rest Energy of Electron, $m_0 c^2 = 0\cdot51$ Mev = $0\cdot8184 \times 10^{-6}$ erg.
Classical Electronic Radius, $r_e = e^2/mc^2 = 2\cdot82 \times 10^{-13}$ cm.

TABLE Ic

Elementary Particles

Particle	Mass	Charge	Mechanical Spin	Magnetic Moment
Neutron	$1{\cdot}6748 \times 10^{-24}$ g.	0	$\tfrac{1}{2}\hbar$	$-1{\cdot}935$ N.M.
Proton	$1{\cdot}6725 \times 10^{-24}$ g.	$1{\cdot}602 \times 10^{-19}$ coul.	$\tfrac{1}{2}\hbar$	$2{\cdot}785$ N.M.
Electron	$9{\cdot}107 \times 10^{-28}$ g.	$-1{\cdot}602 \times 10^{-19}$ coul.	$\tfrac{1}{2}\hbar$	-1 B.M.
Positron	$9{\cdot}107 \times 10^{-28}$ g.	$1{\cdot}602 \times 10^{-19}$ coul.	$\tfrac{1}{2}\hbar$	1 B.M.
Neutrino	~ 0	0	$\tfrac{1}{2}\hbar$	?
Artineutrino	~ 0	0	$\tfrac{1}{2}\hbar$	?
Light Meson	$209\, m_e$	$\pm 1{\cdot}602 \times 10^{-19}$ coul.	not known with certainty.	
π Meson	$276\, m_e$	$\pm 1{\cdot}602 \times 10^{-19}$ coul.		
τ Meson	$\sim 900\, m_e$	$\pm 1{\cdot}602 \times 10^{-19}$ coul.		
Light Quantum (Photon)	0	0	$1 \times \hbar$	0

1 N.M. = 1 Nuclear Magneton = $5{\cdot}505 \times 10^{-24}$ Gauss $\times$ cm^3.

1 B.M. = 1 Bohr Magneton = $9{\cdot}27 \times 10^{-24}$ Gauss $\times$ cm^3 = $1{\cdot}8365$ N.M.

m_e = Mass of Electron.

Note.—The unit of field intensity, the Gauss, is sometimes called the Oersted.

Table II

Chart of the chemical Elements and their Average Chemical Atomic Weights

Element		Atomic Number	Atomic Weight
H	Hydrogen	1	1·0080
He	Helium	2	4·003
Li	Lithium	3	6·940
Be	Beryllium	4	9·02
B	Boron	5	10·82
C	Carbon	6	12·010
N	Nitrogen	7	14·008
O	Oxygen	8	16·0000
F	Fluorine	9	19·00
Ne	Neon	10	20·183
Na	Sodium	11	22·997
Mg	Magnesium	12	24·32
Al	Aluminium	13	26·97
Si	Silicium	14	28·06
P	Phosphorus	15	30·974
S	Sulphur	16	32·066
Cl	Chlorine	17	35·457
A	Argon	18	39·944
K	Potassium	19	39·096
Ca	Calcium	20	40·08
Sc	Scandium	21	45·10
Ti	Titanium	22	47·90
V	Vanadium	23	50·95
Cr	Chromium	24	52·01
Mn	Manganese	25	54·93
Fe	Iron	26	55·85
Co	Cobalt	27	58·94
Ni	Nickel	28	58·69
Cu	Copper	29	63·542
Zn	Zinc	30	65·38
Ga	Gallium	31	69·72
Ge	Germanium	32	72·60

TABLE II (continued)

Element		Atomic Number	Atomic Weight
As	Arsenic	33	74·91
Se	Selenium	34	78·96
Br	Bromine	35	79·916
Kr	Krypton	36	83·7
Rb	Rubidium	37	85·48
Sr	Strontium	38	87·63
Y	Yttrium	39	88·92
Zr	Zirconium, Zircon	40	91·22
Nb	Niobium	41	92·91
Mo	Molybdenum	42	95·95
Tc	Technetium	43	~99
Ru	Ruthenium	44	101·7
Rh	Rhodium	45	012·91
Pd	Palladium	46	106·7
Ag	Silver	47	107·880
Cd	Cadmium	48	112·41
In	Indium	49	114·76
Sn	Tin	50	118·70
Sb	Antimony	51	121·76
Te	Tellurium	52	127·61
I	Iodine	53	126·92
Xe	Xenon	54	131·3
Cs	Caesium	55	132·91
Ba	Barium	56	137·36
La	Lanthanum	57	138·92
Ce	Cerium	58	140·13
Pr	Praseodymium	59	140·92
Nd	Neodymium	60	144·27
—	——	61	—
Sm	Samarium	62	150·38
Eu	Europium	63	152·0
Gd	Gadolinium	64	156·9
Tb	Terbium	65	159·2
Dy	Dysprosium	66	162·46
Ho	Holmium	67	164·94

TABLE II (continued)

Element		Atomic Number	Atomic Weight
Er	Erbium	68	167·2
Tm	Thulium	69	169·4
Yb	Ytterbium	70	173·04
Lm	Lutetium	71	174·99
Hf	Hafnium	72	178·6
Ta	Tantalum	73	180·88
W	Tungsten	74	183·92
Re	Rhenium	75	186·31
Os	Osmium	76	190·2
Ir	Iridium	77	193·1
Pt	Platinum	78	195·23
Au	Gold	79	197·2
Hg	Mercury	80	200·61
Tl	Thallium	81	204·39
Pb	Lead	82	207·21
Bi	Bismuth	83	209·00
Po	Polonium	84	~210
At	Astatine	85	~218
Rn	Radon	86	222
Fr	Francium	87	~223
Ra	Radium	88	226·05
Ac	Actinium	89	~227
Th	Thorium	90	232·12
Pa	Protactinium	91	~231
U	Uranium	92	238·07
Np	Neptunium	93	—
Pu	Plutonium	94	—
Am	Americium	95	—
Cm	Curium	96	—

TABLE III

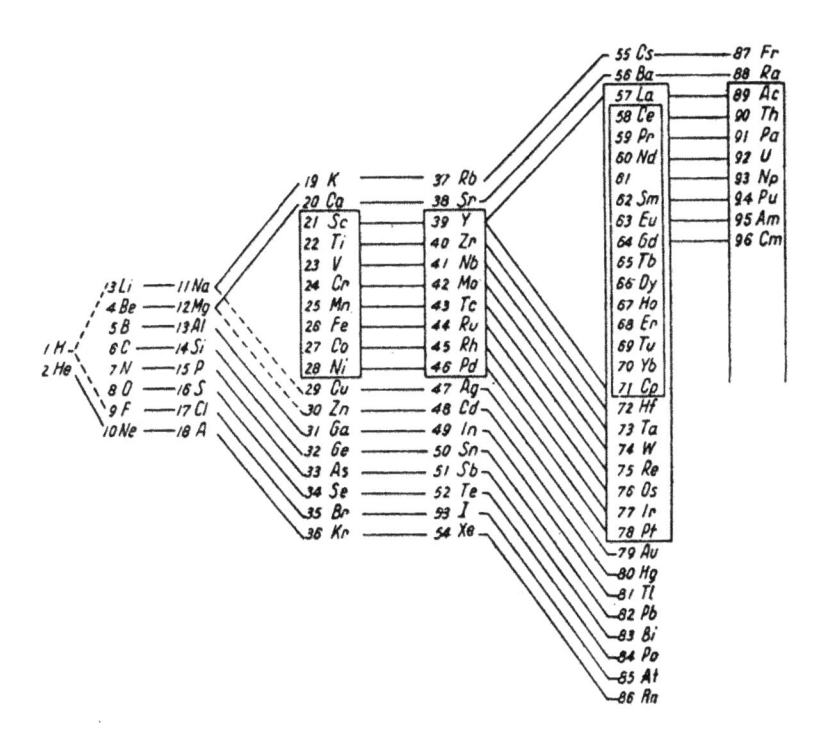

					55 Cs	87 Fr
					56 Ba	88 Ra
					57 La	89 Ac
					58 Ce	90 Th
					59 Pr	91 Pa
					60 Nd	92 U
		19 K	37 Rb	61	93 Np
		20 Ca	38 Sr	62 Sm	94 Pu
		21 Sc	39 Y	63 Eu	95 Am
		22 Ti	40 Zr	64 Gd	96 Cm
		23 V	41 Nb	65 Tb	
3 Li	11 Na	24 Cr	42 Mo	66 Dy	
4 Be	12 Mg	25 Mn	43 Tc	67 Ho	
5 B	13 Al	26 Fe	44 Ru	68 Er	
6 C	14 Si	27 Co	45 Rh	69 Tu	
1 H	7 N	15 P	28 Ni	46 Pd	70 Yb
2 He	8 O	16 S	29 Cu	47 Ag	71 Cp
	9 F	17 Cl	30 Zn	48 Cd	72 Hf
	10 Ne	18 A	31 Ga	49 In	73 Ta
			32 Ge	50 Sn	74 W
			33 As	51 Sb	75 Re
			34 Se	52 Te	76 Os
			35 Br	53 I	77 Ir
			36 Kr	54 Xe	78 Pt
					79 Au
					80 Hg
					81 Tl
					82 Pb
					83 Bi
					84 Po
					85 At
					86 Rn

The periodic system of elements

209

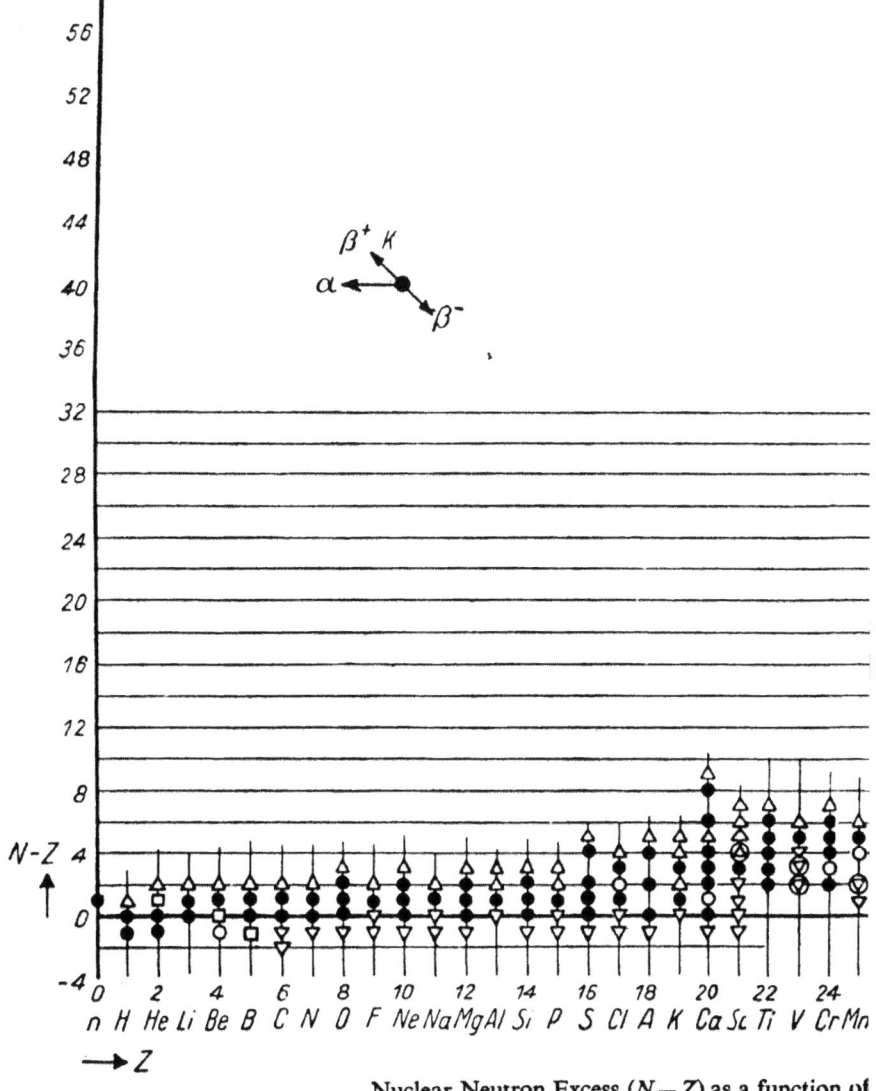

Nuclear Neutron Excess $(N-Z)$ as a function of

210

TABLE IVa

- ● stable nuclei
- △ β⁻-(electron) emitters
- ▽ β⁺-(positron) emitters
- □ α-emitters
- ○ K-electron capturers
- ✡ β-emitters, capable of emitting both electrons and positrons

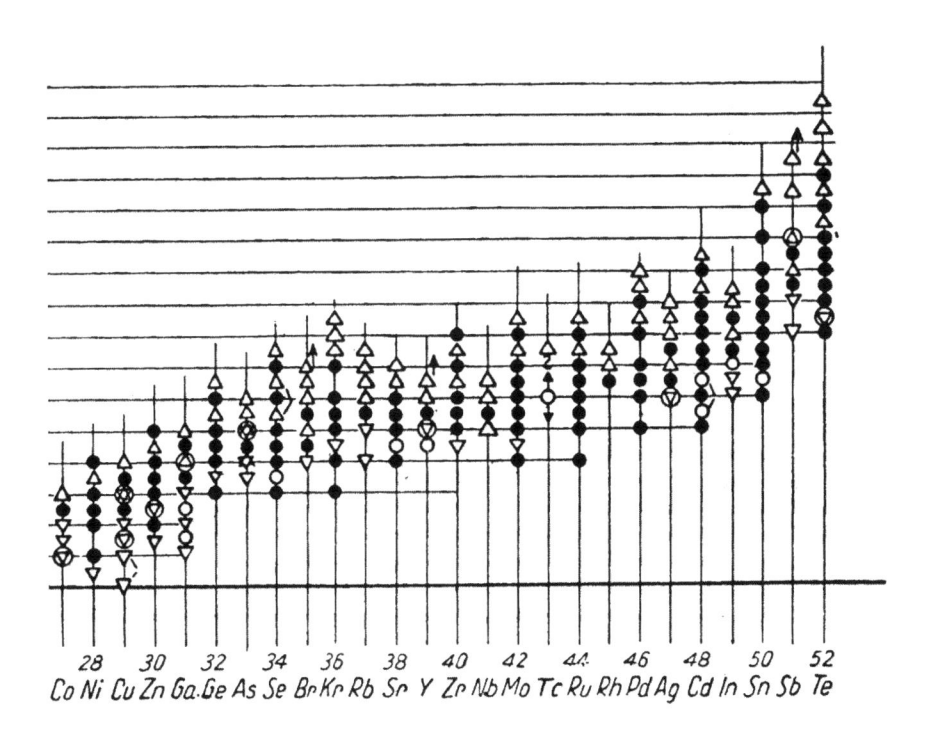

```
 28  30  32  34  36  38  40  42  44  46  48  50  52
 Co Ni Cu Zn Ga Ge As Se Br Kr Rb Sr Y Zr Nb Mo Tc Ru Rh Pd Ag Cd In Sn Sb Te
```

Atomic Number (Z) for the known nuclei.

Nuclear Neutron Excess $(N - Z)$ as a function o

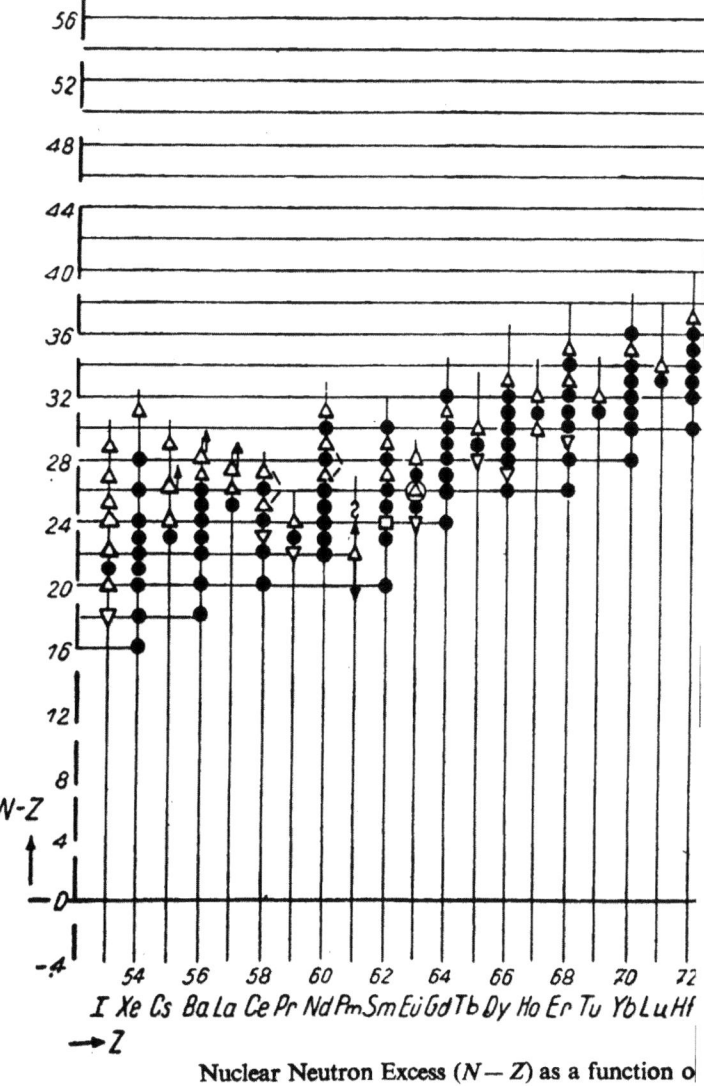

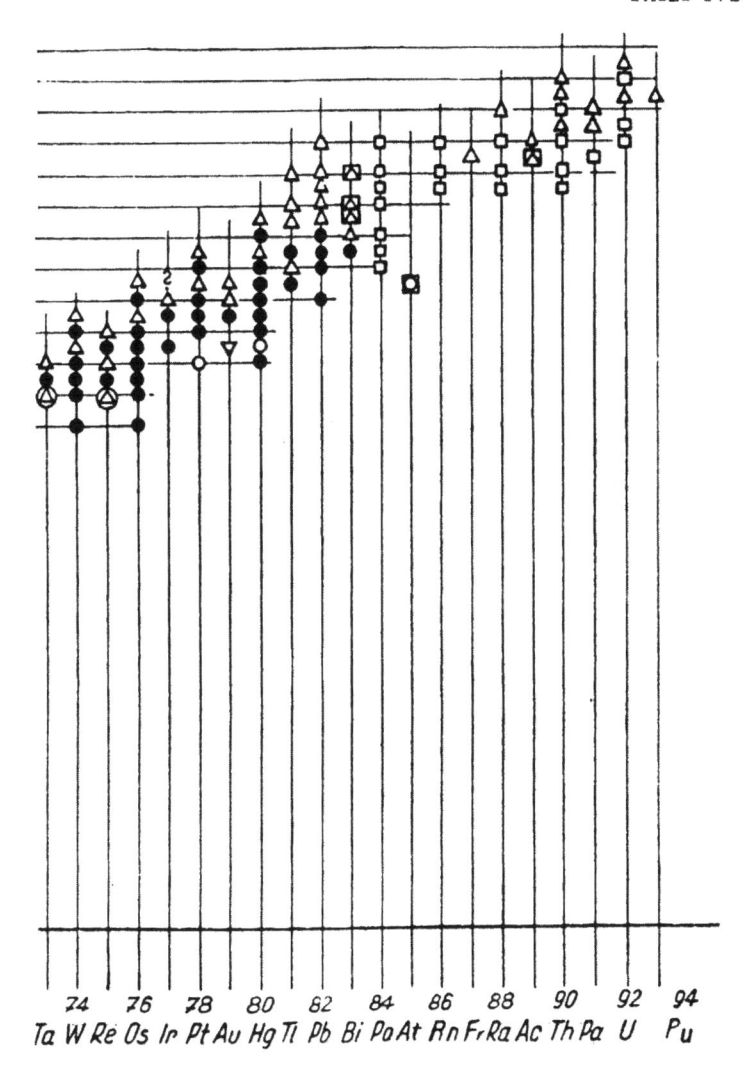

TABLE IVB

74 76 78 80 82 84 86 88 90 92 94
Ta W Re Os Ir Pt Au Hg Tl Pb Bi Po At Rn Fr Ra Ac Th Pa U Pu

ne Atomic Number (Z) for the known nuclei.

Table V

Physical Weights of Isotopes; Abundance and Radiation Characteristics of the Light Elements.[1]

Z = Number of Protons; N = Number of Neutrons in an Atom; T = Half-Life, Relative Abundances indicated in percentages.

Element	Z	N	$Z+N$	Atomic Weight	Relative Abundance	T	Type of Radiation
n	0	1	1	1·008945	—	—	—
H	1	0	1	1·008131	99·985	—	—
D		1	2	2·014725	0·015	—	—
T		2	3	3·017004	—	31 a	β^-
He	2	1	3	3·016988	10^{-5}	—	—
		2	4	4·003860	~100	—	—
		3	5	5·015428	—	~6 × 10^{-20} s	$\alpha + n$
		4	6	6·0209	—	0·8 s	β^-
Li	3	3	6	6·016917	7·9	—	—
		4	7	7·018163	92·1	—	—
		5	8	8·024967	—	0·9 s	β^-, α
Be	4	3	7	7·019089	—	53 d	Kx
		4	8	8·007807	—	< 1 s	2α
		5	9	9·014958	100	—	—
		6	10	10·016622	—	~10^6 a	β^-
B	5	4	9	9·016104	—	unstable	$2\alpha + \mu$
		5	10	10·016169	20	—	—
		6	11	11·012901	80	—	—
		7	12	12·0168	—	0·022 s	β^-
C	6	4	10	10·02086	—	8·8 s	$\beta + \gamma$
		5	11	11·015017	—	21 m	β^+
		6	12	12·003880	98·9	—	—
		7	13	13·007561	1·1	—	—
		8	14	14·007741	—	10^3 to 10^5 a	β^-
N	7	6	13	13·009904	—	9·93 m	$\beta + \gamma$
		7	14	14·007530	99·62	—	—
		8	15	15·004870	0·38	—	—
		9	16	16·00645	—	8·4 s	β

[1] Table V is taken essentially from Mattauch-Flügge's *Kernphysikalische Tabellen*, Berlin 1942. In the statement of half-life values, s stands for seconds, m for months, d for days, and a for years.

TABLE V (continued)

Element	Z	N	Z + N	Atomic Weight	Relative Abundance	T	Type of Radiation
O	8	7	15	15·0078	—	125 s	β^+
		8	16	16·00000	99·76	—	—
		9	17	17·00450	0·04	—	—
		10	18	18·00485	0·20	—	—
		11	19	—	—	31 s	β^-
F	9	8	17	17·00758	—	1·23 m	β^+
		9	18	18·00670	—	107 m	β^+
		10	19	19·00454	100	—	—
		11	20	20·00654	—	12 s	$\beta^-\gamma$
Ne	10	9	19	19·00798	—	20·3 s	—
		10	20	19·998895	90·00	—	β^+
		11	21	21·00002	0·27	—	—
		12	22	21·99858	9·73	—	—
		13	23	23·00084	—	43 s	β^-
Na	11	10	21	—	—	23 s	β^+
		11	22	22·00032	—	3·0 a	$\beta^+\gamma$
		12	23	22·99644	100	—	—
		13	24	23·99774	—	14·8 h	$\beta^-\gamma$
Mg	12	11	23	23·00055	—	11·6 s	β^+
		12	24	23·99300	77·4	—	—
		13	25	24·99462	11·5	—	—
		14	26	25·99012	11·1	—	—
		15	27	26·99256	—	10·0 m	$\beta^-\gamma$
Al	13	13	26	25·99443	—	7·0 s	β^+
		14	27	26·99069	100	—	—
		15	28	27·99077	—	2·3 m	$\beta^-\gamma$
		16	29	28·9892	—	6·7 m	β^-
Si	14	13	27	26·99611	—	4·92 s	β^+
		14	28	27·98723	89·3	—	—
		15	29	28·98651	6·2	—	—
		16	30	29·98399	4·2	—	—
		17	31	30·9866	—	157·3 m	β^-
P	15	14	29	28·99151	—	4·6 s	β^+
		15	30	29·9885	—	130·6 s	β^+

TABLE V (continued)

Element	Z	N	$Z+N$	Atomic Weight	Relative Abundance	T	Type of Radiation
		16	31	30·98441	100	—	—
		17	32	31·98437	—	14·285 d	β^-
		>16	>31	—	—	12·7 s	β^-
S	16	15	31	30·98965	—	3·18 s	β^+
		16	32	31·98252	95·1	—	—
		17	33	32·9819	0·74	—	—
		18	34	33·97981	4·2	—	—
		19	35	—	—	—	—
		20	36	—	0·016	—	—
		21	37	—	—	88 d	β^-
Cl	17	16	33	—	—	2·4 s	β^+
		17	34	—	—	32 m	β^+
		18	35	34·97884	75·4	—	—
		19	36	35·97803	—	(>1 a)	β^+, K, β^-
		20	37	36·97770	24·6	—	—
		21	38	37·97999	—	37·5 m	$\beta^-\ \gamma$
A	18	17	35	—	—	1·88 s	β^+
		18	36	35·97728	0·31	—	—
		20	38	37·97463	0·06	—	—
		22	40	39·97549	99·63	—	—
		23	41	40·97740	—	110 m	$\beta^-\ \gamma$
K	19	19	38	—	—	7·65 m	β^+
		20	39	38·976	93·44	—	—
		21	40	—	0·012	$14·2 . 10^8$ a	β^-
		22	41	—	6·55	—	—
		23	42	—	—	12·4 h	β^-
		24	43	—	—	} 18 m	β^-
		25	44	—	—		
Ca	20	(19)	(39)	—	—	4·5 m	β^+
		20	40	—	96·95	—	
		21	41	—	—	8·5 d	Kγ
		22	42	—	0·64	—	—
		23	43	—	0·15	—	—
		24	44	—	2·07	—	—

TABLE V (continued)

Element	Z	N	Z + N	Atomic Weight	Relative Abundance	T	Type of Radiation
		25	45	44·97075	—	180 d	$\beta^-\ \gamma$
		26	46	—	0·003	—	
		28	48	—	0·185	—	
		29	49	—	—	2·5 h	$\beta^-\ \gamma$
		(isomer)		—	—	30 m	β^-

LITERATURE ON NUCLEAR PHYSICS

I. Short Popular Treatises
1. P. Debye, *Kernphysik*, Leipzig, Hirzel, 1935.
2. L. Meitner and M. Delbrück, *Der Aufbau der Atomkerne*, Springer-Verlag, Berlin, 1935.

II. Introductory Text-books
1. G. Gamow, *Der Bau des Atomkerns und die Radioaktivität*, Hirzel, Leipzig, 1932.
2. W. Riezler, *Einführung in die Kernphysik*, 3rd ed., Bibl. Inst., Leipzig, 1944.
3. F. Rasetti, *Elements of Nuclear Physics*, Blackie and Sons, London, 1937
4. W. Hanle, *Künstliche Radioaktivität*, G. Fischer, Jena, 1939.

III. Monographs on Special Branches of Nuclear Physics
1. C. F. v. Weizsäcker, *Die Atomkerne*, Akad. Verl. Ges., Leipzig, 1937.
2. I. Thibaud, L. Cartanet, P. Comparat, *Quelques techniques actuelles en Physique nucléare*, Gauthiers-Villar, Paris, 1938.
3. H. Kopfermann, *Kernmomente*, Akad. Verl. Ges., Leipzig, 1940.
4. I. Mattauch and S. Flücge, *Kernphysikalische Tabellen*, Springer, Berlin, 1942.
5. H. A. Bethe, *Elementary Nuclear Theory*, John Wiley and Sons, New York, 1947.
6. L. Rosenfeld, *Nuclear Forces*, North Holland Publ. Comp., Amsterdam, 1948.
7. O. Hahn, *Künstliche neue Elemente*, Verlag Chemie, Wiesbaden, 1948.
8. H. Dänzer, *Einführung in die theoretische Kernphysik*, Braun, Karlsruhe, 1949.
9. W. Bothe and S. Flügge, *Kernphysik und kosmische Strahlung*, Naturf.u.Med. in Deutschland, Vol. 13 u. 14, Dieterische Verl.-Buchh., Wiesbaden, 1949.
10. W. Heisenberg and K. Wirtz, *Untersuchungen zum Atomenergieproblem*, Dieterische Verl.-Buchh., Wiesbaden, 1949:
11. H. D. W. Smyth, *Atomenergie und ihre Verwendung im Kriege*, Reinhardt, Basel. 1947.

ACKNOWLEDGMENTS FOR ILLUSTRATIONS

2. Grimsehl's *Lehrbuch der Physik*, 7.Aufl. revised by Thomaschak, Vol. 1, Fig. 305, p. 264. Verlag G. B. Teubner, Leipzig.
3. Meitner-Freitag, *ZS. f. Phys.* 37, 481, 1926, *Table I* (*S*.634), *Ill.* 2
5 and 6. *Die physikalischen Prinzipien der Quantentheroie* by W. Heisenberg, Ill. 1, Table 1, Verlag S. Hirzel, Leipzig.
9, 10 and 28. *Atlas typischer Nebelkammerbilder* by Gentner, Maier-Leibnitz, Bothe, Ill. 26 a/b, p. 74; Ill. 30, p. 78; Ill. 17, 0. 40, Springer-Verlag, Berlin.
17A. Original Photograph, by courtesy of Mr. Lawrence.
25A and 25B, C. F. Powell and G. P. S. Occhialini, *Nuclear Physics in Photographs*, Clarendon Press, Oxford, 1947.
27. *Die Naturwissenschaften*, Vol. 21, p. 477, Ill. 3, Year 1933, *Atomzertrümmerung durch Wasserstoffkanalstrahlen* by Kirchner, Springer-Verlag, Berlin.
34, 35, 36, 37 and 38. *Elektrische Höchstspannungen* by A. Bouwers, Ill. 46, p. 52; Ill. 49, p. 58; Ill. 53, p. 62; Ill. 56, p. 68; Ill. 60, p. 71, Springer-Verlag, Berlin.

INDEX

Actinium series, 114
Alchemists' dream, 187
Alchemy, 6, 139
Alpha, beta and gamma rays, 24, 42, 172
Alpha particles, 43, 54
— —, energy and range of, 114
— radiation, 113
Americium, 181
Anaxagoras, 2
Anaximander, 2
Anderson, 56
Angular momentum, 55
Annihilation radiation, 48
Antineutrino, 55
Antiquity, philosophy of, 1
Argon, 40
Aristotle, 5
Astatine, 180
Aston, 73
Atkinson, 161
Atom, extranuclear structure of, 33
—, ground state of, 36
—, size of, 12, 13, 17, 18
Atomic bombs, 167, 201
— energy, utilization of, 190
— mass unit, 72
— nucleus, 42
— number, 38
— structure of electricity, 11
— theory, 5
— theory, history of, 14
— theory, in antiquity, 1
— theory, modern, 4
— weight, 9
— weight, unit of, 21
Atoms, 3
Atoms and the void, 3
— of electricity, 13
—, shape of, 14
—, weights of, 14
Avogadro, 9
—'s hypothesis, 9, 19
—'s hypothesis, proof of, 19

Bagge, 56
Becquerel, Henri, 16

Being and becoming, 2
— — not being, 2, 3
Beryllium, 40
Berzelius, 9
Beta radiation, 122
Bethe, 161
Biermann, 56
Binding energy, of atomic nuclei, 66, 67
— —, per particle, 83
Black body, 31
— — radiation, 31
Blackett, P., 52
Blau, 132
Bohr, N., 31, 32, 38, 128, 191
— magneton, 55
Boltzmann, L., 12
Boron, 40
— and carbon, 126
— counter, 147
Born, 177, 182
Bothe, 53, 130, 190, 193, 195, 197, 198
Boyle, Robert, 6, 7, 8
—'s law, 6
Broglie, Louis de, 32
— waves, 35
Brownian movement, 16
Butler, 56

Cadmium, as a moderator, 169
Carbon, 40, 194
— as a catalyst, 163
— dioxide, assimilation of, 183
Cathode rays, 13
Chadwick, Sir J., 53
Chain reaction, 164, 166, 169, 189, 192
Chalmers, 175
Clausius, 12
Cloud chamber, 25, 52, 146
— — tracks, 147
Cockcroft, Sir J., 137
Condensation nuclei, 25
Condon, 116
Conservation of mass, 8
Constant proportions, law of, 8

Corpuscles, red and white, 184
Corson, McKenzie and Segré, 180
Cosmic radiation, 55, 132, 161
— —, primary, 56
Coulomb's law, 27
Counter, 143, 144, 145
Curie, 24, 53
Curium, 28, 36, 181
Cyclotron, 99, 153

Dalton, John, 8, 9
—, atomic theory of, 9
Debye, P., 191
Decay probability, 45, 124
Dees, 155
Democritus, 3, 4, 5, 7, 10
Detection of neutrons, 147
Deuterium, 63, 186
Deuteron, 63, 67
—, binding energy of, 70
—, mass of, 74
Diebner, 191
Dirac, P., 48
Döpel, 169, 193
Dynamides, 26

Earth's surface, 161
Ehmert, 56
Einstein, 70
—'s mass-energy law, 70, 71
Electricity, atom of, 22
—, atomic structure of, 11
Electrolysis, 11
Electron capture, 126
— emitters, 64, 87, 122
— microscope, 18
— volt, 69
Electronic shells, 39
Electrons, 13, 14, 57
—, mass of, 13
—, planetary, 27
Electroscope, 141
Elements and compounds, 7, 10
—, chemical, 7
—, transmutation of, 42, 43, 44
Empedocles, 3
Epicureans, 4
Equivalent weights, 11
Erbacher, 174, 177
Exchange forces, 96, 98, 99, 103
Exclusion principle, 39, 105
Exothermic process, 165

Extranuclear structure of atom, 58, 61
— — — — and electric field, 58
— — — — and photons, 58

Faraday, M., 11, 13
Fermi, 130, 131, 170
Field and particles, 97
Fine structure, 187
Fission, 89, 134, 165
—, in U^{238}, 198
Flügge, 189, 190
— and v. Droste, 83
Fluorine, 40
Forbush, 56
Francium, 181
Fundamental building blocks, 60

Gamma rays, 46, 47, 173
Gamov, 116
Gases, internal friction of, 12
—, theory of, 12
Gassendi, 5, 8
Geiger, 27
Geiger counter, 143
Geiger–Müller counter, 144
Geiger–Nuttall law, 115, 121
Gerlach, 200.
Gramme-atom, 12
Greinacher circuit, 150
Gurney, 116

Hahn, 173, 174, 190
Hahn and Strassmann, 89, 127, 134, 189, 202
Half-life of radium, 115
— of radium C′, 114
— of uranium, 114
— period, 45
Halogens, 41
Harkins, 106
Hasenöhrl, 70
Heavy meson, 56
— water, 168, 192, 194
Heisenberg, W., 192, 193, 197
Helium, 38, 39, 63
— atom, 28
—, formation of, 162, 163
— nuclei, 43, 114
Heraclitus, 2
Hess and Kohlhörster, 55
Hevesy, 182, 183

INDEX

Hittorf, 13
Houtermans, 161
Hydrogen, 8, 38, 40
— atom, electron orbits of, 36, 37
— —, model of, 28
Hydrogen fluoride, 41
— molecule, mass of, 22

Illinium, 180
Inert gases, 39, 40
Ionization chamber, 141
Isobars, nuclear, 86
Isotopes, 62, 75, 111, 112
— of mercury, 187

Joliot, 53, 166, 189
Joliot-Curie, 137

Karlik and Bernert, 180
K capture and neutrino emission, 127
K capturers, 88, 108
K shell, 88
K radiation, 127
Kirchhoff, 31
Kirchner, 137
Krypton, 40

Lavoisier, 7, 8
Law of constant proportions, 8
Lawrence, 99, 153
Lenard, 26
Leprince-Ringuet, 5, 6
Leucippus, 3
Liquid drop model of nucleus, 79
Lithium, 39
—, electropositive character of, 40
Loschmidt, 12, 13, 21
—'s number, 21

Marsden, 27
Mass, conservation of, 8
— defects, 73
— of deuteron, 74
— ratio, 20
Mass spectrograph, 73
Masurium, 178
Mattauch's rule, 110
Matter, ennoblement of, 172
Maxwell, 12, 19, 90
—'s equations, 90

Mayer, Robert, 12
Meitner and Frisch, 189
Meson, 55, 56, 102
Mesothorium, 173
Metabolism, 182, 183
Miletus, 1
Moderator, 168
Mole, 12, 21
Molecule of water, 8, 15, 16, 20
Molecules and atoms, 15
—, kinetic energy of, 19
—, structure of, 15, 20
Müller, 144

Neon, 40
Neptunium, 166
Neutrino, 50, 76, 102
Neutron, 53, 57
— excess, 65
— injected piles, 193
Neutrons, detection of, 147
—, spectrum of, 193
Nietzsche, Friedrich, 1
Nitrogen, 40
Nuclear charge, 27
— "combustion", 163
— cross-section, 131
— energy, 67
— —, types of, 80
— field, 60, 61
— fission, 134
— forces, 90, 94
— forces, saturation of, 103, 104
— isobars, 86, 111
— magneton, 55
— physics, 1
— physics, constants of, 23, 204, 205
— —, practical applications of, 159
— reaction, 54, 59
— reactions in sun and stars, 161
— structure, 77
— transmutations, 113, 149, 179
— — in the atmosphere, 56
Nuclei, binding energy of, 66
Nucleus, 42
—, liquid drop model of, 79
Nucleus, stability of, 75, 104

Oppenheimer, 167
Oxygen, 7, 8, 40

Pair formation, 48
Parmenides, 2
Pauli, 39, 50
—, principle of, 39, 105
Periodic system, 38
Perrey, 181
Philipp, 177
Phlogiston, 7
Phosphorus, radioactive, 174
Photon, 37
Planck, 21, 31, 123
—'s constant, 32
—'s radiation law, 21, 31
Plato, 4
—'s *Timaeus*, 4
Plutonium, 165, 166, 194
Point counter, 143
Positive electron, 23
— ions, 25
Positron emitters, 64, 87, 122
Positrons, 47, 57
Potential barrier, 95, 118, 120
— container, 118
Powell, 102, 132
Probability value, 35
Proportional count, 143
— region, 143
Proton, 51, 57
—s, attraction between, 95
Prout, 10, 11, 52, 60, 74, 112
Pythagoras, 4

Quantum mechanics, 32, 55
— theory, 30

Radiation, black body, 31
—, cosmic, 55
—, nuclear, 113
Radioactive decay, 45
— phosphorus, 174
— series, 114
Radioactivity, 24, 42, 113
—, as a label, 175
Radium, 24, 44, 173
Range of alpha particles, 24, 45, 49
Rays, alpha, beta and gamma, 24
Resolving region, 144
Rochester, 56
Rock salt, 17
Röntgen, Wilhelm, 23
Ruben, Hassid and Kamen, 183

Rutherford, Ernest, 2, 4, 26, 42, 43, 51, 127, 136
Rutherford's atom model, 23, 27, 29, 31

Sargent diagram, 125
Schopper, 132
Schrödinger, E., 35, 36
Schumann, 191
Scintillation method, 141
Scott and Cook, 184
Soddy, 42, 43
Sodium, 41
Space, 3
Spin, 55, 105
Stoney, 13
Stopping distance, 198
String electrometer, 142
Surface tension in atomic nuclei, 79
Szilard, 17

Telephone counter, 145
Thales, 1, 2
Thorium series, 114
Thyratron, 145
Tracer, 148, 175, 177, 182
Transmutation of nitrogen into oxygen, 51
Tritium, 63
Triton, 63
Tunnel effect, 121
Types of nuclear energy, 80

Uncertainty principle, 30, 93
Unsöld, 56
Uranium nucleus, diameter of, 78
— pile, 170, 171
— reactor, 167, 168, 169
— series, 114
—, (U^{235}), 165, 195, 199
—, (U^{238}), 165
—, (U^{239}), 194
—, X, 126

Valency, 9. 93
—, electrical nature of, 10
— forces, 96
Van de Graaff, 152
Volume ratios, 20

Walton, 137
Wambacher, 132
Water molecule, 8, 19
Wave aspect of light, 58
Wave mechanics, 32
Wave-particle duality, 32, 97, 98, 121
Weber, 11
Weizsäcker, 80, 161, 193
Williams, W. E., 187

Wilson, C. T. R., 25, 146

Xenon, 40
X-rays, 23, 33

Yukawa, 101

Zero point vibration, 95
Zimmer, 177

9 780806 530338

Printed in Great Britain
by Amazon